CHRISTMAS IN THE CORPS

Holiday Stories & Poetry By and About Marines

Compiled by MSgt A. A. Bufalo USMC (Ret)

ISBN   978-0-9845957-0-9

First Printing – October 2010
Printed in the United States of America

www.AllAmericanBooks.com

OTHER BOOKS BY ANDY BUFALO

SWIFT, SILENT & SURROUNDED
Sea Stories and Politically Incorrect Common Sense

THE OLDER WE GET, THE BETTER WE WERE
MORE Sea Stories and Politically Incorrect Common Sense

NOT AS LEAN, NOT AS MEAN, STILL A MARINE
Even MORE Sea Stories and Politically Incorrect Common Sense

**EVERY DAY IS A HOLIDAY,
EVERY MEAL IS A FEAST**
A Fourth Book of Sea Stories and Politically Incorrect Common Sense

THE ONLY EASY DAY WAS YESTERDAY
Fighting the War on Terrorism

TO ERR IS HUMAN, TO FORGIVE DIVINE
However, Neither is Marine Corps Policy
A Book of Marine Corps Humor

HARD CORPS
Legends of the Marine Corps

SALTY LANGUAGE
An Unabridged Dictionary of Marine Corps Slang, Terms &Jargon

THE LORE OF THE CORPS
Quotations By, About & For Marines

**"I'll be home for Christmas,
You can count on me.
I'll be home for Christmas,
If only in my dreams."**

– Bing Crosby

FOR

Major Gene "Dunk" Gene Duncan

and

All Marines who are in harm's way or separated from their loved ones during the Holidays.

PREFACE

Christmas in the Corps is a labor of love which was compiled in response to a request from some Marine Parents whose sons were deployed to Iraq during the holidays. A few of the stories were drawn from my previous books, and the rest from a variety of sources, with the common thread being the service of Marines during the Holiday season. Most of these stories are heartwarming, but there are also a couple which remind us of the sacrifices of military service and the realities of war.

Major Gene Duncan once wrote that otherwise tough Marines are "suckers for kids and dogs," and that is evident in the three dog stories in this volume. Also interspersed throughout the holiday tales are three "Marine-ized" versions of Clement C. Moore's immortal *The Night Before Christmas.*

The cover design was inspired by a Christmas card produced by the Marine Corps Association which shows a Marine arriving home on a snowy Christmas Eve. It is of course impossible for them all to come home for the Holidays, even during peacetime, so it is my hope they can all be safe while enjoying the camaraderie of their fellow Marines. Happy Holidays, and Semper Fi!

TABLE OF CONTENTS

Christmas in the Corps

10

CHRISTMAS DAY
All Secure!

'Twas the night before Christmas, he lived all alone,
In a one-bedroom house made of plaster and stone.
I had come down the chimney with presents to give,
And to see just who in this home did live.

I looked all about, a strange sight I did see,
No tinsel, no presents, not even a tree.
No stocking by mantle, just boots filled with sand,
On the wall hung pictures of far distant lands.

With medals and badges, awards of all kinds,
A sober thought came through my mind.
For this house was different, it was dark and so dreary,
I found the home of a Marine, once I could see clearly.

The Marine lay sleeping, silent, alone,
Curled up on the floor in this one bedroom home.
The face was so gentle, the room so unclean,
Not how I pictured a United States Marine.

Was this the hero of whom I'd just read?
Curled up on a poncho, the floor for a bed?
I realized the families that I saw this night,
Owed their lives to these Marines, so willing to fight.

Soon 'round the world, the children would play,
And grownups would celebrate a bright Christmas Day.
They all enjoyed freedom each month of the year,
Because of the Marines, like the one lying here.

I couldn't help wonder how many lay alone,
On a cold Christmas Eve in a land far from home.
The very thought brought a tear to my eye,
I dropped to my knees and started to cry.

Christmas in the Corps

The Marine awakened and I heard a rough voice,
"Santa don't cry, this life is my choice.
I fight for freedom, I don't ask for more,
My life is my God, my country, my Corps."

The Marine rolled over and drifted to sleep,
I couldn't control it, I continued to weep.
I kept watch for hours, so silent and still,
And we both shivered in the night's cold chill.

I didn't want to leave on that cold, dark night,
This guardian of honor so willing to fight.
Then the Marine rolled over, with a voice soft and pure,
Whispered, "Carry on Santa, it's Christmas Day, all secure."
One look at my watch, and I knew he was right.
"Merry Christmas my friend, and to all a good night."

A CHRISTMAS STORY

Major H. G. Duncan

A collection of characters lived in our Staff NCO tent when I was a staff sergeant with the 1st Battalion, 5th Marines in Korea. There was 'Elias,' the German. There was 'Smock,' who fancied himself an intellectual and stayed busy counting down his last ten days in Korea. There was 'Cowlick,' who was bitter at the world after an unhappy marriage and three wayward kids. There was 'Geoffrey.' Geff and I constituted what Elias called the '"children's corner" of the tent. We were both young, and looked it, and had to endure the verbal abuse of the senior citizens of our tent.

Geff had befriended an old Korean man who worked at the mess tents as a pot-wholloper, and announced on Christmas Eve morning that the old man wanted to visit our tent that afternoon to wish us a Merry Christmas. Cowlick looked at Geff, sneered, lit a cigarette and left the tent. Elias asked "Vat duz dat kook vant?" I was indifferent. Smock was busy staring at his calendar.

We were all in the tent that afternoon at about three o'clock when Geff ushered in the old man and his family, which consisted of his daughter and her three children. One of the small ones was obviously blind. Another had only one leg. The third appeared to be a perfectly healthy little girl of about four years.

Geff had told me earlier that the old man had lost his wife to the war and had moved in with his daughter and her family. Her husband had long since been killed while serving in the Korean Army. An enemy artillery round had landed close to their house during the initial invasion of South Korea, and had blinded the boy and amputated his sister's leg.

The old man, who spoke fairly good English, told us his family wanted to come by and give their Christmas greetings to "the Malines who fight for Korea." He then said something in the native tongue to his flock, and they lined up and began singing in Korean.

To this day I've never heard a more beautiful *Silent Night* or *Joy to the World.*

The thing that fascinated me about this little group was their perpetual smiles. Each of them smiled genuinely and continuously.

After the singing they presented each of us with a small handmade gift, obviously crafted as a labor of love for people they hardly knew.

I looked at Cowlick. His disinterest had turned to a look of confused bewilderment. Smock had put his calendar aside, and sat on his rack looking at the kids. Elias was smiling back at the little one-legged girl. Geff was in his Christian glory. I felt a kind of heavy heart, but with buoyant spirits, if that makes any sense.

Elias excused himself, left the tent, and returned shortly with some stolen baked goods from the mess tent. We then broke out some carefully guarded cokes and liquor.

Our little group then had a small "tea" with this Korean family, mixing hot chocolate for the kids which they drank from our seemingly huge canteen cups.

Cowlick motioned the little girl with one leg over to his packing crate chair. She hobbled over on her crutches, and he lifted her to his lap. The old man's smile grew wider.

The little blind boy sat on Elias' lap. Elias attempted to give the kid several dollars in our 'scrip' money, but the old man politely – but firmly – intervened and would not permit it. He then took one of his dogtags off his necklace and gave it to the boy. The old man nodded his approval.

I thought for a moment and then reached into my rolled up sleeping bag and withdrew my treasured bottle of bourbon. I handed it to the old man who smiled, placed his hands in front of him as a shield, and shook his head. I insisted, and he finally accepted.

Geff then gave the young mother a gift he had obviously purchased for her. It was a new kettle.

We all sat and stood there in awkward silence. Cowlick was the first to speak. In his gruff but somehow gentle voice he said, "The Staff NCO choir will now sing for you." Then to us, "Gather around, you clowns."

We sang the English versions of *Silent Night* and *Joy to the World*. We got most of the words wrong, but the Koreans never noticed. They were now smiling more brightly than ever.

It was soon time for them to go. Geff escorted the family to the main gate, where they started the five mile trek back to their bombed out, patched up house.

Christmas had worked its never-failing magic on us. We sat around in silence for awhile, until Cowlick spoke up. "Elias, if you ever catch me feeling sorry for myself again, I want you to kick me right square in the ass."

FOUR WISE MEN
From the East

Captain W. T. Moore

With only four more days until Christmas, thoughts have a habit of returning to less troubled times. On the 21st of December in 1951 I graduated from Marine Corps boot camp at MCRD San Diego, and along with seventy-eight other newly minted Marines I packed my seabag and boarded the bus to the train station or airport.

We divided into groups destined for large

geographical areas. The train riders who formed my group were bound for the Midwest and points east. We climbed aboard the Santa Fe Super Chief, and I estimated we made up the majority of the passengers in two complete coach cars.

During boot camp I had formed 'everlasting' friendships that have since evaporated into nothing more than pleasant memories. In the car to which I was assigned were several of my best buddies, including Mike Happney of Wolf Lake, Michigan, Jimmy Phillips of Saint Paul, Minnesota, and Fred Delaney from Saint Louis, Missouri. The four of us seated ourselves in a pair of coach seats which had been turned so that they faced each other. As a result, two members of our foursome were obliged to ride facing the rear of our train.

Our gear was stowed, uniform coats removed, covers placed in the racks above, and the train pulled out of the station eastbound. I glanced around and was proud to observe that every passenger in the coach was a Marine with the exception of three persons.

Across the aisle, riding in a double seat arrangement similar to ours, was a family – a dark-complected young man of about twenty-five with a wife who could have been a high fashion model and a daughter of about five years who looked like a doll. This family appeared as though they were dressed for a part in *The Grapes of Wrath*. They were not dirty, but had obviously seen hard times.

The coach was filled with the sounds of Marine boots full of the joy of graduation, the Christmas spirit, and going home! The little family concealed their self-consciousness by looking out the window.

Mike Happney pulled out his harmonica and Fred Delaney broke out a jug of Four Roses. It wasn't long before several bottles of cheer appeared and were seen moving up and down the aisle. As Mike played Christmas carols on the mouth organ, the Marines started to sing. I noticed the little girl across the aisle had joined in several of the songs. I couldn't hear her voice, but I could see her mouth moving and it was obvious she knew the words to the more popular songs. Jimmy motioned to the father and offered him a drink. It was accepted. We introduced ourselves and learned that the family was enroute to Oklahoma to visit the wife's mother who would not live to see another Christmas. The father, who introduced himself as Alvin, told us unashamedly that he and his wife had saved all they could for this trip, and it became apparent to us that he had used all he had and would be unable to provide much of a Christmas for his little girl.

The Super Chief rolled on through the night and the Christmas party accelerated. The little girl was now outfitted in a Marine garrison cap and was seated between Fred and Jimmy singing with gusto. Several of the older Marines had made the long trip to the club car and produced great quantities of cold beer and soda. The lone civilian family had been adopted by about eighty Marines. From somewhere back in the club car came a small Christmas tree with lights which were battery powered – several strings of red, silver and green lights with like-colored tinsel. The tree was set up on a suitcase in the center of the aisle. The bright tinsel and bells were strung the length of the car, and an atmosphere of goodwill prevailed.

Soon the train had crossed the width of California, the great Arizona desert and the wilds of New Mexico. Sleep had come to many of us, and thus refreshed the party was joined again. The train wound its way across the great Southwest, and at various points Marines departed the mobile Christmas party to join family and friends – and a few empty seats began to appear in the coach. These empty seats were soon occupied by Marines who put them to use in the prone position, either passed out from excess partying or just plain tired after the long trip.

The train pulled to a stop in San Antonio, Texas, and the conductor informed us that we would have a half-hour to make phone calls, stretch our legs, or pick up some souvenir Texas dirt. About sixty-five Marines bailed out of that car. The station, as I recall, looked like pictures I had seen in my high school history book of the Alamo. Across the street from the depot was a large drug or department store. My fire team – Jimmy, Fred, Mike and I – headed across the street and found the toy department. When we reentered the train station each of us had a large stuffed animal and, of course, Fred was armed with another fifth of Four Roses.

As we approached the Marines who were milling about the waiting room, several asked where we had obtained the dolls and teddy bears. Happney yelled, "Attention! All you guys form up over here!" He gathered a crowd of at least fifty Marines. "Okay, listen up! There's a little girl riding in our car who is going to Oklahoma to see her grandma, and grandma is fixing to check out before the kid has another Christmas with her. Her old man ain't got a crying dime, and that kid ain't gonna have much of a Christmas. So I'm going to let you clowns in on a hell of a deal. Follow me over to that big store across the street, and I'll get you a special buy on a present for the kid. Fall in!"

So there was Mike Happney with a platoon of Marines, with road guards out, crossing the street and marching in right through the front door of that big store. He brought them to a halt and called for the manager. Mike said to the guy, "If I tell these Marines to, they will buy every goddam toy you got in this place. How about a fifteen percent discount?"

The manager allowed as to that was a hell of a deal, if for no other reason it was a good way to rid himself of fifty half drunk Marines who were eyeing his female customers in a way that made even the old ones feel uneasy. With less than fifteen minutes to departure time those guys grabbed every kind of stuffed animal, doll and toy designed for a young girl that was on the shelves. There was no time for gift wrapping or cards, so the platoon formed up on the sidewalk outside the store and, with Happney in charge, marched back to the train.

As each Marine entered the coach and passed the double seat occupied by Alvin and his family he blurted out, "Merry Christmas!" and deposited his gift for the little girl.

Well, it wasn't long before that girl was entirely covered with animals and dolls, so much so that she was completely out of sight. Her mother was crying, and so was Alvin – and, as I recall, there were a few Marines seen wiping away a tear or two. The conductor was summoned to the car, and several 'stiff' Marines were rearranged and transferred to other seats. Two more seats were turned to form a fore-and-aft arrangement, and the loot was deposited so that the little girl would be able to survive the remainder of the trip.

The Christmas party was renewed with vigor, and I've never heard *Silent Night* done better – even by a church choir.

As the Super Chief pulled into the station in that small Oklahoma town you could feel that special chill that touches all of us at that time of year. The conductor had several Marines and porters carry off the gifts and put them on one of those high four wheel push carts that were used for baggage. Alvin put his daughter on the cart right in the middle of that huge pile of toys and pushed her right past the windows of the car so she could wave goodbye to the trainload of Marines. As she stood smiling and waving, the cart stopped in front of our window. Happney glanced out at her, and then with a slight wave he looked at Delaney and said, "Gimme a drink. Let's get the hell out of here."

After all, we were a bunch of tough Marines!

THE GUNNY SACK BRIDE

MSgt Dennis De Noi USMC (Ret)

It was the night before Christmas in 1991, and my ship was off the Korean peninsula in a winter storm. Deep in the warship's belly, seventy-five young Marines sat in cramped quarters in the detachment's multifunction area reading holiday greetings sent to "Any Sailor, *USS Independence*."

Mail call, the most important event of the day at sea, was more exciting due to the influx of thousands of cards from patriotic Americans responding to a notice in a "Dear Abby" column inviting citizens to write military personnel during the holidays. The Gulf War had ended, but the ensuing campaigns to ensure stability in that fragile region continued.

The ship's mailroom, already being overrun with holiday packages, now had mountains of gunnysacks filled with Dear Abby mail arriving as well. The mailroom personnel urgently processed the holiday cheer by stuffing hundreds of cards into the ship's mail slots. Mail calls were doubled to ensure the crew did not lose the mail-processing battle. Back at the Marine Detachment, the guard chief (me) watched as the young men, many away from home for the first time, opened letter after letter.

They laughed, joked and exchanged the contents of the unsolicited mail – but it was their actions that followed that disturbed me. After reading the cards, they just threw many of them in the trash without replying. I knew why these Americans had taken the time to write us, and felt strongly that we owed them at least one response. I called a formation, and laid out new rules for the Dear Sailor holiday mail: From that point, any Marine who opened a letter *would* reply and tell the originator about the joy the letter brought and a little about life on the ship.

The young warriors responded in typical Marine Corps fashion, and what followed was quite amazing to watch. The letters kept coming, and those who wanted to participate continued to open and respond to the holiday greetings. Having decided the letters were intended for the young Marines, I derived my joy from watching them – but that was about to change, as was my life forever.

Late one night, as I sat at my desk inside the detachment area, a young Marine knocked on my door. He was holding a large envelope containing about thirty letters and drawings from second graders in California. He said he had opened and read the contents, but felt one letter was better suited for me and he wanted to give it to me so I could respond.

I thanked him and explained that the program was for the young Marines, and that is when he got me. He said that along with the kids' letters was a card explaining the reason the kids had written, and how important it was for them to receive a reply. The kicker was the "P.S." The teacher, Cheryl Tucker, added that she was a thirty-year-old single parent, and she wanted to know if the recipient would like to become her pen pal. The nineteen-year-old Marine quickly arrived at the answer: No! But how would he get out of the obligation to write back?

I'm still not quite sure how this kid half my age, and whom I greatly out-ranked, won the debate, but when he left the office I was still holding the envelope. With my own words echoing in my head, I decided to respond. After all, I was also a thirty-something single parent without a solid lead on a soul mate with whom to enjoy my life and share my dreams. I responded with a letter to the thirty new charges in my life by grouping their names according to their questions, drew pictures that further enhanced my responses, and enclosed a card of my own to the adventurous teacher.

Thus began an eighteen-month courtship of cards, letters and photographs which eventually led to my transfer to Southern California upon completion of sea duty. My pen pal met me at the Los Angeles Airport USO on the last day of school, and we had our first date in the company of her parents at a quaint restaurant. We were married a year later, and have been best friends ever since.

I continue to visit the school as the longest-running school project in the district's history. Without the Marine Corps, and the holiday wishes of patriotic Americans, I would not have the happy and fulfilling life I now enjoy with my Christmas gift/mail order bride from the States!

THE CHRISTMAS COOKIES

Arnold I. Pakula

In late 1953, while preparing for Christmas services at the chapel of Marine Corps Air Station, Opa-Locka, Florida (then home of the 3rd Marine Air Wing) I created a major faux pas that set Judeo-Christian relations back to the time of Christians vs. lions.

It was the week before Christmas, and my duties as chaplain's assistant included overseeing the completion of the decoration of the chapel by volunteer Navy and Marine Corps wives in preparation for Christmas services.

However, I first had to complete typing up the Catholic chaplain's Christmas sermon for his final approval, and while doing so decided to have some of the coffee that it had been my job to brew. After pouring myself a mug, and as I typed the sermon, I had an overwhelming urge for some cookies to go with my coffee. I usually kept a full box in my desk in anticipation of such urges, but on this particular morning it was empty. I decided to go looking in the supply closet beside the chapel, since I knew from past experience there was always something interesting in there.

As soon as I opened the closet door I spied what appeared to be an open box of cookies. Although there was no brand name on the box, and even though they had no creamy middles, I decided "any cookies in a storm."

As soon as I placed the first one in my mouth it seemed to instantly evaporate, and it had no cookie taste at all. Then I attempted to dunk one into my mug of hot coffee, and it disappeared almost instantly. It took no time at all to go through the entire box, but I was left feeling as if I'd had no cookies at all.

Later, as I was finishing up the sermon with the empty box sitting on my desk beside my now empty coffee mug, the Catholic chaplain walked past my office, waved, did a double-take, and backpeddled to my desk – where he stared with unbelieving eyes at the empty box. Without his even asking, I innocently explained my desperate need for cookies with my coffee. He then calmly explained that what I had just eaten where not cookies at all. They were Communion wafers! After swallowing hard, I wondered just how many wafers it took to change a young Jewish boy into a practicing Catholic.

It was my good fortune the chaplain had a good sense of humor. As he walked out of the office, clutching tightly to his newly typed Christmas sermon in one hand and the empty Communion wafer box in the other, his deep-throated laughter followed after him.

Nothing more was said of the incident until the following Sunday morning, when I was mentioned in the opening remarks of his sermon – bringing laughter from the entire congregation of Navy and Marine personnel and their families. The only repercussion came with the New Year and my amicable transfer to the office of the Commanding Officer, Headquarters Squadron, Marine Wing Service group 37... along with a box of cream-filled cookies.

COMING HOME

This is a different kind of coming home story.

Jerome Lee of Stonewall has been a trooper for Troop H of the Mississippi Highway Patrol since 1982. He can still vividly remember his six-year old son, Dustin, playing in his patrol car.

"He always wanted to call the dispatcher on the radio to tell them I was in service," said Jerome. "I let him play with the siren and lights some. He wanted to become a state trooper."

But Dustin Jerome Lee won't be able to fulfill his dream of being a trooper and following in his father's footsteps. Dustin, age twenty, died serving as a Marine Corps in Iraq – just six weeks before he was expected to return home.

The Lee family was notified of their sacrifice by Marine Corps officials.

"He was a very focused, intense individual," said the elder Lee. "He always excelled in everything he set out to do. He had his whole life mapped out."

Dustin was serving in 3rd platoon G Battery 3/14, in Fallujah when he was killed in a mortar attack.

"He was hit in the chest with shrapnel from the blast and was medevaced out of the area to a hospital," said Jerome. "He died a little while later."

The two men talked the day before Dustin was killed. Jerome said his son was sounding upbeat and in high spirits.

"Maybe it was the knowledge he was coming home in about six weeks. I don't know. But he left me a voice message on my answering machine before he went out on his last mission." In the message Dustin said he just wanted to call before he headed out to tell everyone he loved them. "He said he'd talk to us later," Jerome said.

But like so many things we wish for in life, somehow that longed-for tomorrow never came.

Dustin's Dad remembers his son with pride. "He always wanted to help other people. He loved his country, and was proud to be a Marine."

Asked, despite the devastating loss, if he was proud of his son's service to his country and his sacrifice Jerome replied quietly, "Most definitely. I'm very proud of him."

Some families would have turned inward in their grief, would have shut out the world and thought only of their own pain. But the Lees, in the midst of their own heartbreak, somehow found the grace to remember that they weren't the only ones mourning the loss of their son.

Marine Corporal Dustin Jerome Lee and his German shepherd, Lex, had scoured Iraq for roadside bombs together, slept next to each other, and even posed in Santa hats for a holiday photo. When a mortar attack killed the young Marine in Fallujah a few months later Lex, whimpering from his own injuries, had to be pulled away.

That strong bond compelled the slain Marine's family to petition the Corps to adopt the eight-year-old dog, even though the military said he still had two years of service. The family lobbied the military for months, launched an Internet petition, and enlisted the aid of a North Carolina congressman who took their case straight to the Marine Corps' top general - but the Marine Corps still had doubts. Lex had recovered from his wounds and was still fit for duty, bomb sniffing dogs were in short supply in the Corps, and they save lives. Would sending Lex home be the right thing to do?

"I know Dustin would want Lex to be with his family," said Lee's uncle, Brian Rich. "They gave their son - he made the ultimate sacrifice. If it brings his family some comfort to see the dog there, then why not?"

The Lees took their case to the Commandant of the Marine Corps, and on December 21st their persistence was rewarded when they finally took Lex home for the holidays.

They were always together: the tall, handsome young Marine from Mississippi, and the German Shepherd specially trained to help him find roadside bombs - but in the end, that wasn't the most important thing. The most important thing was they were both Marines, and when you're a Marine you always have family - no matter what.

I'll be home for Christmas... if only in my dreams...

THE EMBERS GLOWED SOFTLY

The embers glowed softly, and in their dim light,
I gazed round the room and I cherished the sight.
My wife was asleep, her head on my chest,
My daughter beside me, angelic in rest.

Outside the snow fell, a blanket of white,
Transforming the yard to a winter delight.
The sparkling lights in the tree, I believe,
Completed the magic that was Christmas Eve.

My eyelids were heavy, my breathing was deep,
Secure and surrounded by love I would sleep.
In perfect contentment, or so it would seem,
So I slumbered, perhaps, I started to dream.

The sound wasn't loud, and it wasn't too near,
But I opened my eyes when it tickled my ear.
Perhaps just a cough, I didn't quite know,
Then the sure sound of footsteps outside in the snow.

My soul gave a tremble, I struggled to hear,
And I crept to the door just to see who was near.
Standing out in the cold and the dark of the night,
A lone figure stood, his face weary and tight.

A soldier, I puzzled, some twenty years old,
Perhaps a Marine, huddled here in the cold.
Alone in the dark, he looked up and smiled,
Standing watch over me, and my wife and my child.

"What are you doing?" I asked without fear,
"Come in this moment, it's freezing out here!
Put down your pack, brush the snow from your sleeve,
You should be at home on a cold Christmas Eve!"

For barely a moment I saw his eyes shift,
Away from the cold and the snow blown in drifts,
To the window that danced with a warm fire's light
Then he sighed and he said, "It's really all right,
I'm out here by choice. I'm here every night.

It's my duty to stand at the front of the line,
That separates you from the darkest of times.
No one had to ask or beg or implore me,
I'm proud to stand here like my fathers before me.

My Gramps died at Pearl on a day in December."
Then sighed, "That's a Christmas Gram always remembers."
My dad stood his watch in the jungles of 'Nam
And now it's my turn, and so here I am.

I've not seen my own son in more than a while,
But my wife sends me pictures, he's sure got her smile.
Then he bent and he carefully pulled from his bag,
The red white and blue... an American flag.

"I can live through the cold and the being alone,
Away from my family, my house and my home,
I can stand at my post through the rain and the sleet,
I can sleep in a foxhole with little to eat.

I can carry the weight of killing another,
Or lay down my life with my sister and brother.
Who stand at the front against any and all,
To ensure for all time that this flag will not fall."

"So go back inside," he said, "harbor no fright,
Your family is waiting and I'll be all right."
"But isn't there something I can do, at the least,
Give you money," I asked, "or prepare you a feast?
It seems all too little for all that you've done,
For being away from your wife and your son."

Then his eye welled a tear that held no regret,
"Just tell us you love us, and never forget.
To fight for our rights back at home while we're gone,
To stand your own watch, no matter how long.

For when we come home, either standing or dead,
To know you remember we fought and we bled.
Is payment enough, and with that we will trust,
That we mattered to you as you mattered to us.

A MAN I NEVER KNEW

Lynn Ashby

It was a setting right out of a Christmas card. The snow was gently falling on the ground in lazy white flakes. The night air was crisp and quiet. The surrounding hills stood white against the black sky. Ahead was a small village with food and fires. Thomas R. Connelly of Houston and his traveling companions moved down the road with a quickened pace, for it was Christmas Eve, and a more idyllic yuletide scene could not have been imagined.

But the situation was not exactly as it seemed. The year was Christmas 1950 - a time when there was no peace on Earth and no good will toward men, particularly in North Korea somewhere above the Chosin Reservoir. And Connelly was no holiday carouser making merry in a winter wonderland. Rather he was a staff sergeant in the U.S. Marine Corps, a combat veteran of World War II, and now a prisoner of war, one of one hundred and fifty American soldiers and Marines being moved north to a POW camp.

With the invasion of South Korea, American forces from all over had been thrown into the defense of the peninsula. Most were ill-prepared for combat. U.S. Army troops had been enjoying soft garrison duty in Japan and were out of shape and often poorly led. I clearly remember an old Marine telling me that on the ship from Japan to Korea a soldier was practicing firing his weapon into the ocean while asking, "How do you shoot this thing?" As for the Marines, the Marine Corps had been reduced after World War II to a small shadow of its previous strength. So, in combat, U.S. troops were overwhelmed. Many were killed or captured, which brings us to this sorry situation.

Connelly's group was heavily guarded by North Korean soldiers, but the Americans were not particularly dangerous. They had been force marched for days, tied together in a long ugly line of frostbitten feet and ragged uniforms. They had not eaten in three days, and many in the group were wounded. "It was dark, probably somewhere between ten PM and one AM," Connelly recalled. "I don't know exactly, because the North Koreans took our watches. But it had been dark for some time."

With much pushing and prodding, the POWs staggered down the road. "As we entered the small village, we were stopped and made to kneel in the middle of the road. Our guards started a search of the village for food for themselves, and to get warm by the fire."

As the prisoners waited in the freezing night the villagers timidly appeared, one by one.

"The people came slowly out to look us over. They seemed to be only a crowd of expressionless faces. They neither smiled nor spoke. Only solemn stares seemed to meet our eyes. Finally, a little girl emerged from the crowd. She was about seven or eight years old, it was hard to tell. She watched us for a while, and then slowly approached the young

Marine in front of me. She smiled and reached out and handed him a small handful of rice. I couldn't tell whether it was cooked rice or what, but the Marine started eating it."

She said in broken English, "Christian." The young Marine smiled at her and said, "Merry Christmas." She looked puzzled at first, and then her eyes lit up, and she said, "Christmas is Jesus." Her English was not very good, but he could tell that she spoke it with an American accent, not British. Apparently she had been taught English by American missionaries. She never knew about Santa Claus, or brightly lighted Christmas trees. She probably never knew that on Christmas morning she was to receive a gift, or really knew or had even heard the many Christmas carols we sing.

"Christmas is Jesus," she said again. At this very moment one of the guards, seeing that she had passed something to one of us, screamed at her in Korean. You could see the fear in her eyes and, raising her hand, she showed the guard she had only given some rice. But for some unknown reason - I never knew why - the guard raised his rifle and fired at her. Quickly, we were forced to our feet and moved out of the village. In silence, the one hundred and fifty Americans shuffled past the small figure lying in the reddening snow.

"I believe it was one of the greatest gifts ever given on Christmas," said Connelly. "Of all the prayers and carols that will be sung at Christmas, none will ever say, 'Christmas' as much as that little girl. She gave a small gift of love and said simply, "Christmas is Jesus."

A few years later I tried to find Thomas Connelly again, but never could.

THE LIBERTY LIMITED

It started last Christmas, when Bennett and Vivian Levin were overwhelmed by sadness while listening to radio reports of injured American troops. "We have to let them know we care," Vivian told Bennett. So they organized a trip to bring soldiers from Walter Reed Army Medical Center and Bethesda Naval Hospital to the annual Army-Navy football game in Philadelphia on December 3rd – and the cool part is, they created their own train line to do it!

Yes, there are people in this country who actually

own real trains. Bennett Levin - native Philly guy, self-made millionaire and irascible former L&I commissioner - is one of them. He has three luxury rail cars. Think mahogany paneling, plush seating and white-linen dining areas. He also has two locomotives, which he stores at his Juniata Park train yard. One car, the elegant *Pennsylvania*, carried John F. Kennedy to the Army-Navy game in 1961 and '62. Later, it carried his brother Bobby's body to D.C. for burial. "That's a lot of history for one car," said Bennett.

He and Vivian wanted to revive a tradition that endured from 1936 to 1975, during which trains carried Army-Navy spectators from around the country directly to the stadium where the annual game is played. The Levins could think of no better passengers to reinstate the ceremonial ride than the wounded men and women recovering at Walter Reed in D.C. and Bethesda in Maryland. "We wanted to give them a first-class experience. Gourmet meals on board, private transportation from the train to the stadium, perfect seats - real hero treatment."

Through the Army War College Foundation, of which he is a trustee, Bennett met with Walter

Reed's Commanding General, who loved the idea. But Bennett had some ground rules first, all designed to keep the focus on the troops alone. No press on the trip, lest the soldiers' day of pampering devolve into a media circus. No politicians either, because, said Bennett, "I didn't want some idiot making this trip into a campaign photo op." And no Pentagon 'suits' on-board, otherwise the soldiers would be too busy saluting superiors to relax. The general agreed to the conditions, and Bennett realized he had a problem on his hands. "I had to actually make this thing happen," he laughed.

Over the next months he recruited the owners of fifteen other sumptuous rail cars from around the country - these people tend to know each other - into lending their vehicles for the day. The name of their temporary train? *The Liberty Limited*. Amtrak volunteered to transport the cars to D.C. where they'd be coupled together for the round-trip ride to Philly, and then bring them back to their owners later. Conrail offered to service the *Liberty* while it was in Philly, and SEPTA drivers would bus the disabled soldiers the two hundred yards from the train to Lincoln Financial Field for the game. A benefactor from the War College ponied up one hundred seats to the game - on the fifty-yard line - and lunch in a hospitality suite, and corporate donors filled, for free and without asking for publicity, goodie bags for attendees. From Woolrich, stadium blankets. From Wal-Mart, digital cameras. From Nikon, field glasses. From GEAR, down jackets. There was booty not just for the soldiers, but for their guests too, since each was allowed to bring a friend or family member. The Marines, though, declined the offer. "They voted not to take guests with them, so they could take more Marines," said Levin, choking up at the memory.

Bennett is an emotional guy, so he was worried about how he'd react to meeting the eighty-eight troops and guests at D.C.'s Union Station, where the trip originated. Some were missing limbs. Others were wheelchair-bound or accompanied by medical personnel for the day. "They made it easy to be with them," he said. "They were all smiles on the ride to Philly. Not an ounce of self-pity from any of them. They're so full of life and determination." At the stadium, the troops reveled in the game. Not even Army's lopsided loss to Navy could deflate the group's rollicking mood. Afterward, it was back to the train and yet another gourmet meal - heroes get hungry, after all - before returning to Walter Reed and Bethesda. "The day was spectacular," said Levin. "It was all about these kids. It was awesome to be part of it."

The most poignant moment for the Levins was when eleven Marines hugged them goodbye, and then sang the Marines' Hymn on the platform at Union Station for them. "One of the guys was blind, but he said, 'I can't see you, but man, you must be f---ing beautiful!'" said Bennett. "I got a lump so big in my throat, I couldn't even answer him."

Three weeks later the Levins and their guests were still feeling the day's love. "My Christmas came early," said Levin, who is Jewish and loves the Christmas season. "I can't describe the feeling in the air."

Maybe it was hope. As one guest wrote in a thank-you note to Bennett and Vivian, "The fond memories generated last Saturday will sustain us all - whatever the future may bring."

God bless the Levins. And bless the troops, every one.

SILENT NIGHT

Marine boot camp can be a frightening place, but a very funny place as well. Here's one of my favorite Christmas boot camp stories as experienced by one of my buddies.

It was Christmas Eve night and the recruits were out in the boondocks as part of their training - but the drill instructors weren't even acknowledging the occasion. Finally, one of the recruits summoned up the courage to broach the issue.

"So you want Christmas, huh? Okay then, let the games begin!" said the DI.

The DI then proceeded to have half the platoon climb a big tree with their flashlights so that the recruits were evenly spaced throughout the branches. Half of these recruits were ordered to put the red lens in their flashlights, while the others kept the white lens in. Then the DI ordered the recruits in the tree to sing *Silent Night* while switching on and off their flashlights.

As such, the half of the platoon that didn't climb the tree were witness to a real live singing Christmas tree complete with blinking red and white lights. After the song was over the recruits in the tree switched places with those on the ground and repeated the experience, so that everyone had a very Marine Christmas that they will never forget.

THE PAINTING

Years ago there was a very wealthy man who, along with his devoted young son, shared a passion for art collecting. Together they traveled around the world, adding only the finest art treasures to their collection. Priceless works by Picasso, Van Gogh, Monet and many others adorned the walls of the family estate. The widowed elder man looked on with satisfaction as his only child became an experienced art collector. The son's trained eye and sharp business mind caused his father to beam with pride as they dealt with art collectors around the

56

world.

As winter approached, war engulfed the nation and the young man left to serve his country in the Marine Corps. After only a few short weeks, his father received a telegram. His beloved son was missing in action. The art collector anxiously awaited more news, fearing he would never see his son again. Within days, his fears were confirmed. The young man had died while carrying a fellow Marine across a fireswept battlefield to a Corpsman. Distraught and lonely, the old man faced the upcoming Christmas holidays with anguish and sadness. The joy of the season - a season that he and his son had always looked forward to - would visit his house no longer.

On Christmas morning a knock on the door awakened the depressed old man. As he walked through the large house to the door, the masterpieces of art on the walls only reminded him that his son was not coming home. He opened the door, and was greeted by a young Marine with a large package in his hand. He introduced himself to the man by saying, "I was a friend of your son. I was the one he was rescuing when he died. May I come in for a few moments? I have something to show you."

As the two began to talk, the Marine told of how the man's son had told him of his father's love of fine art. "I'm an artist myself," said the Marine, "and I want to give you this." As the old man unwrapped the package, the paper gave way to reveal a portrait of his son in his Dress Blues. Though the world would never consider it the work of a genius, the painting featured the young man's face in striking detail. Overcome with emotion, the man thanked the Marine, promising to hang the picture above the fireplace. A few hours later, after the Marine had departed, the old man set about his task. True to his word the painting went above the fireplace, pushing aside many thousands of dollars worth of paintings. He then sat in his chair and spent Christmas gazing at the gift he had been given.

During the days and weeks that followed the man realized that even though his son was no longer with him, the boy's life would live on because of those he had touched. He would soon learn that his son had rescued dozens of wounded Marines before a bullet stilled his caring heart. As the stories of his son's gallantry continued to reach him, fatherly pride and satisfaction began to ease the grief. The painting of his son soon became his most prized possession, far eclipsing any interest in the pieces for which museums around the world clamored. He told his neighbors it was the greatest gift he had ever received.

The following spring the old man became ill and passed away. The art world was in anticipation. With the collector's passing, and his only son dead, those paintings would be sold at an auction. According to the will of the old man all of the art works would be auctioned on Christmas Day, the day he had received his greatest gift. The day soon arrived, and art collectors from around the world gathered to bid on some of the world's most spectacular paintings. Dreams would be fulfilled this day and greatness would be achieved as many would claim, "I have the greatest collection." But the auction began with a painting that was not on any museum's list. It was the painting of the man's son. The auctioneer asked for an opening bid. The room was silent.

"Who will open the bidding with one hundred dollars?" he asked. Minutes passed. No one spoke.

From the back of the room came, "Who cares about that painting? It's just a picture of his son.

Let's forget it and go on to the good stuff." More voices echoed in agreement.

"No, we have to sell this one first," replied the auctioneer. "Now, who will take the son?"

Finally, a friend of the old man spoke. "Will you take ten dollars for the painting? That's all I have. I knew the boy, so I'd like to have it."

"I have ten dollars. Will anyone go higher?" called the auctioneer. After more silence, the auctioneer said, "Going once, going twice. Gone." The gavel fell.

Cheers filled the room and someone exclaimed, "Now we can get on with it and bid on these treasures!"

After a few moments the auctioneer looked at the audience and announced the auction was over. Stunned disbelief quieted the room. Someone spoke up and asked, "What do you mean it's over? We didn't come here for a picture of some old guy's son. What about all of these paintings? There are millions of dollars of art here! I demand you explain!"

The auctioneer replied, "It's very simple. According to the will of the father, whoever takes the son... gets it all."

HOME FOR CHRISTMAS

John McCain

Nothing crushes your spirit more effectively than solitary confinement. Having no one else to rely on, to share confidences with, to seek counsel from, you begin to doubt your judgment and your courage. The loneliness robs you of everything - everything but time. When you are in solitary confinement you have nothing to think about other than time and just making it through another day. So needless to say, keeping track of the date is not difficult for a man held at length in solitary confinement.

In the five and a half years I was a prisoner of war in Vietnam, Christmas was always the most difficult time of year for me. I distinctly remember Christmas Eve 1969. I had been a POW for more than two years already, most of which was spent alone in my cell. Like many other cells in the Hanoi Hilton, mine was a small, empty room, roughly seven feet by ten feet with a concrete slab on the floor which served as my bed. The walls were eighteen inched thick, and the windows of each cell were boarded up so that the POWs could not communicate with each other. I remember there being a single, naked lightbulb dangling on a cord in the center of the ceiling, and a small loudspeaker in the corner on which the Vietnamese would play various propaganda pieces.

It was about eight o'clock on Christmas Eve 1969. I was in pretty bad shape, having received some severe beatings from the North Vietnamese. On top of that, I had still not recovered from the injuries I received when I was shot down two years earlier. I was cold. I was injured. And as I lay there in my cell listening to Hanoi Hanna report on "the latest heroic victory over the American imperialists," I had some real serious doubts about my chances for survival.

Then the prison guards began to play a series of Christmas songs over the camp's public address system, the last of which was Dinah Shore singing *I'll Be Home for Christmas*. As I lay there listening to that particular song, my spirits dropped to the lowest possible point. I was not sure if I would survive another night, let alone ever return home for another Christmas with my family.

It was then that I heard the tapping on my wall. Despite the strict rule against it, the POWs communicated to each other by rapping on the walls of our cells. The secretive tap code was a simple system. We divided the alphabet into five columns of five letters each. The letter K was dropped. A, F, L, Q and V were the key letters. Simply tap once for the five letters in the A column, twice for F, three times for L, and so on. After indicating the column, pause for a beat, then tap one to five times to indicate the right letter. For example, the letter C is sent as: *tap... tap tap tap.*

We became so proficient at the tap code that in time the whole prison system became a complex information network. With each new addition to our population, word quickly passed from cell to cell about every POW's circumstances and information from home. The tap code was my sanity's saving grace. That daily personal contact through the drumming on my wall made my isolation more bearable. It affirmed my humanity and kept me alive.

The cell on one side of me was empty, but in the other adjacent room was a guy named Ernie Brace. Ernie was a decorated former Marine who had flown more than one hundred combat mission in the Korean War. He had volunteered as a civilian pilot to fly missions to secretly supply CIA-supported military units in the Laotian jungle. During one such operation in 1965 he was captured and handed over to the North Vietnamese. He was brutally tortured and kept in solitary confinement for three years at a remote outpost near Dien Bien Phu before he was even brought to the Hanoi Hilton in 1968.

As soon as I heard the tapping on Christmas Eve, I knew it was Ernie. I got up and pressed my ear against the cold stone wall of my cell. At first it was difficult to make out the faint tapping of my neighbor. But it soon became very clear.

"We'll all be home for Christmas," Ernie tapped. "God bless America."

With that I began to cry.

When you are imprisoned, the enemy can take almost everything from you but they cannot take your spirit. Those unspoken words coming from Ernie - who, due to his work with the CIA, had the least chance of getting out of the camp alive - were a poignant affirmation that as Americans, we possessed a divine spark that our enemies could not extinguish - hope.

"We'll all be home for Christmas. God bless America."

That simple message, in my darkest hour, strengthened my will to live. Ernie helped me realize that we would get home when we got home. Until then, we had to manage our hardships as best we could. Without his strength, I doubt I would have survived solitary confinement with my mind and self-respect intact.

It was long ago and far away. But around the holidays, when I hear "I'll Be Home for Christmas," I am always reminded of that time, that place, and the words of my friend Ernie Brace. He kept me going and lifted my spirits when they were in their greatest need of lifting. When I hear that song I think about Ernie. I think about my friends that never made it home for another Christmas. And I think of what a blessing it is to be an American.

TOYS FOR TOTS

Toys for Tots began in 1947 when Major Bill Hendricks, USMCR and a group of Marine Reservists in Los Angeles collected and distributed 5,000 toys to needy children. The idea came from Bill's wife, Diane. In the fall of 1947 Diane handcrafted a Raggedy Ann doll and asked Bill to deliver it to an organization which gave toys to needy children at Christmas. When Bill determined that no such agency existed, Diane told him that he should start one. He did. The 1947 campaign was so successful that the Marine Corps adopted Toys for

Tots in 1948 and expanded it into a nationwide campaign. That year Marine Corps Reserve units across the nation conducted Toys for Tots campaigns in each community in which a Reserve Center was located, and Marines have conducted successful nationwide campaigns at Christmas each year since then.

Bill Hendricks, a Marine Reservist on weekends, was in civilian life the Director of Public Relations for Warner Brothers Studio. This enabled him to convince a vast array of celebrities to provide their support. In 1948 Walt Disney designed the Toys for Tots logo, which is still used today. Disney also designed the first Toys for Tots poster used to promote the nationwide program. Nat "King" Cole, Peggy Lee, and Vic Damone recorded the Toys for Tots theme in 1956, and Bob Hope, John Wayne, Doris Day, Tim Allen and Kenny Rogers are but a few of the long list of celebrities who have given their time and talent to promote this worthy charity. First Lady Barbara Bush served as the national spokesperson in 1992, and in her autobiography named Toys for Tots as one of her favorite charities.

My own affiliation with the Toys for Tots program began during my days with 2nd Battalion,

25th Marines in Garden City, New York. Collection points were set up at Shea Stadium prior to a Jets football game one Christmas season, and volunteers were needed to man them. I admit my motive wasn't completely noble that day – participants were allowed into the stadium to watch the second half for free. But the experience did teach me a lot about the spirit of giving.

Once I returned to active duty, deployments and operational tempo often limited my participation in the program - but I still donated toys whenever possible. When I reported for duty in the Congo during the later part of 1992 one of the first things that struck me was the overwhelming poverty of the Congolese people. I wanted to help in some small way, so my Marines placed a large box under the Embassy Christmas Tree and sent a memo to every office soliciting the donation of toys. Everyone was very generous, and what we collected was distributed to local children by missionaries working in the area.

My next assignment was in Australia. While there was nothing like the poverty I saw in the Congo, there were still a lot of needy kids all the same. Once again a collection point was set up under the

Embassy Christmas Tree, and once again we were overwhelmed by the generosity of everyone who worked there. Since there was no official Toys for Tots distribution system there we turned the toys over to the Australian Salvation Army, and they in turn brought them to children's hospitals where the neediest children received an unexpected visit from Santa Claus.

No story about Toys for Tots would be complete without mentioning the late Sam Dipoto, who ran the program for the Marine Corps League in the Tampa/St. Petersburg area of Florida for nearly twenty years. Sam loved to tell the story of a woman who donated a dozen brand new bicycles one Christmas. When he tried to thank her she said it was just *her* way of thanking Toys for Tots. It turned out a couple of years earlier the woman had been out of work and destitute, but the Marines had come through with a bicycle for her young son.

I have no doubt that there are many like Sam all across the country, and hopefully there will be many more to follow in their footsteps. The kids are counting on it.

YULETIDE MEMORIES

The holiday season is often a difficult time of year for those in the military. I have walked a post on guard duty on many a holiday, and have been deployed for many more. I was even in boot camp for my first Christmas in uniform. I especially remember one season while I was stationed in Okinawa. As the holidays approached I recall browsing through the PX down at Camp Foster. There was Christmas music playing in the background, and it was quite pleasant until the song changed to Bing Crosby's "I'll Be Home for

Christmas." A woman shopping nearby went berserk, and began to scream hysterically that she was NOT going to be home for Christmas for the second straight year. It was quite a scene.

On Christmas Eve those of us who could not afford to fly back to the States for the holidays, or who had to remain behind to hold down the fort, headed out the gate of Camp Hansen in search of some sort of diversion. With so many people gone on leave "Sinville" was pretty dead, and we ended up in a Japanese bar singing karaoke in a drunken stupor, which wouldn't have been so bad except they only had one song in English and we sang it over, and over, and over. To this day I can't hear "Yesterday" by the Beatles without cringing. Such wonderful Yuletide memories!

As Christmas approaches this year with Marines still serving in Iraq, Afghanistan and elsewhere, I thought it would be appropriate to include the true story behind a poem many of us have seen circulating for years, along with the inspiring text of that poem.

This piece, which sees wide circulation every Christmastime, is often credited to "a Marine stationed in Okinawa, Japan" (or, since 9-11, a

Marine stationed in Afghanistan). Unfortunately, an Air Force Lieutenant Colonel named Bruce Lovely once took credit for composing this poem, claiming he penned it on Christmas Eve 1993 while stationed in Korea - and he saw it printed under his name in the *Fort Leavenworth Lamp* a few years later:

"I arrived in Korea in July 93 and was extremely impressed with the commitment of the soldiers I worked with and those that were prepared to give their lives to maintain the freedom of South Korea. To honor them, I wrote the poem and went around on Christmas Eve and put it under the doors of U.S. soldiers assigned to Yongsan."

Regrettably, Lieutenant Colonel Lovely did a great disservice to his fellow servicemen by claiming authorship of "The Soldier's Night Before Christmas," because it had already been published in *Leatherneck Magazine* in December 1991, a full two years before Lovely supposedly wrote it. The *Leatherneck* version was titled "Merry Christmas, My Friend" and was attributed to James M. Schmidt, then a Lance Corporal stationed at Marine Barrack 8th & I in Washington, D.C.

According to Corporal Schmidt:

"The true story is that while I was a Lance Corporal serving as Battalion Counter Sniper at the Marine Barracks 8th & I, Washington, DC, I wrote this poem to hang on the door of the gym in the BEQ. When Colonel Myers came upon it, he read it and immediately had copies sent to each department at the Barracks and promptly dismissed the entire Battalion early for Christmas leave. The poem was placed that day in the Marine Corps Gazette, distributed worldwide, and later submitted to Leatherneck Magazine."

The following is Corporal Schmidt's version as printed in *Leatherneck*, which differs from the current Internet version in many places (particularly in Marine-specific wording that has since turned into Army references, and alterations in other places to maintain the line-ending rhyme scheme).

MERRY CHRISTMAS
My Friend

'Twas the night before Christmas, he lived all alone,
In a one-bedroom house made of plaster and stone.
I had come down the chimney, with presents to give,
And to see just who in this home did live.

As I looked all about, a strange sight I did see,
No tinsel, no presents, not even a tree.
No stocking by the fire, just boots filled with sand.
On the wall hung pictures of a far distant land.

Christmas in the Corps

With medals and badges, awards of all kind,
A sobering thought soon came to my mind.
For this house was different, unlike any I'd seen.
This was the home of a U.S. Marine.

I'd heard stories about them, I had to see more,
So I walked down the hall and pushed open the door.
And there he lay sleeping, silent, alone,
Curled up on the floor in his one-bedroom home.

He seemed so gentle, his face so serene,
Not how I had pictured a U.S. Marine.
Was this the hero, of whom I'd just read?
Curled up in his poncho, a floor for his bed?

His head was clean-shaven, his weathered face tan.
I soon understood, this was more than a man.
For I realized the families that I had seen on that night,
Owed their lives to these men, who were willing to fight.

Soon around the Nation, the children would play,
And grown-ups would celebrate on a bright Christmas Day.
They all enjoyed freedom, each month and all year,
Because of Marines like this one lying here.

Christmas in the Corps

I couldn't help wonder how many lay alone,
On a cold Christmas Eve, in a land far from home.
Just the very thought brought a tear to my eye.
I dropped to my knees and I started to cry.

He must have awoken, for I heard a rough voice,
"Santa, don't cry, this life is my choice.
I fight for freedom, I don't ask for more,
My life is my God, my country, my Corps."

With that he rolled over, drifted off into sleep,
I couldn't control it, I continued to weep.
I watched him for hours, so silent and still,
I noticed he shivered from the cold night's chill.

So I took off my jacket, the one made of red,
And covered this Marine from his toes to his head.
Then I put on his T-shirt of scarlet and gold,
With an Eagle, Globe and Anchor emblazoned so bold.

And although it barely fit me, I began to swell with pride,
And for one shining moment, I was a Marine deep inside.
I didn't want to leave him so quiet in the night,
This guardian of honor, so willing to fight.

But half asleep he rolled over, and in a voice clean and pure,
Said "Carry on, Santa, it's Christmas Day, all secure!"
One look at my watch and I knew he was right,
Merry Christmas my friend, Semper Fi and goodnight!

IF I COULD JUST
Remember His Name ...

Paul Scimone

I've never encountered a veteran from 'Nam who had total and clear recollection of the names of guys he served with for over a year. It seems to be the most common disease shared by nearly all Vietnam vets. Several personal incidents remain embedded in my mind as clearly as if they had happened only yesterday, and yet I can't recall one single name.

There is one name in particular I wish I could remember, because I sat up with him all night waiting for him to die, *praying* for him to die, wanting him to be at peace. Maybe I was wishing it for myself so I could go back to my hooch, close my eyes, and dream about being back home with my family and friends.

As the company driver/litter-bearer for 3rd Medical Battalion triage in Phu Bai from 9/67 to 9/68 my duties included driving the ambulance, taking the docs (corpsmen and surgeons) wherever they needed to go, transporting the KIA's to graves registration, etc. Seeing dead, twisted and torn young men was a daily occurrence, and it was somehow easy to become complacent and detached about what you saw and how you felt. After all... "better him than me," right?

On Christmas Eve of 1967 things were relatively quiet around our area. I can remember only one medevac coming in around 9 PM with just one casualty: a 2/26 Marine who'd been shot "through and through" with a head-shot. It didn't take long for the docs to determine that he was brain-dead, even though they were amazed at his almost normal vitals. The docs instructed me to wheel him back to S&D (a shock and debridement room) on a gurney, from where I was sure I'd be taking him to Graves the next morning. Just leave him in that dark cool room all night, and he'll slip away quietly and peacefully. After all, he couldn't feel anything... wasn't aware of anyone or anything around him... so what's the difference?

I went back to my hooch and BS'd with my buds. We sat around quietly that Christmas Eve night talking about family, girlfriends and wives, what we were going to do when we got home, and stuff like that. Everyone took a turn at telling their teary-eyed stories and having a good cry with each other.

I don't know if it was that I was so drunk, had the holiday blues, or what, but I found myself wandering back over to a pretty quiet and dark triage and feeling compelled to check in on the young Marine in S&D. I suddenly felt compelled to stay with him. It was Christmas Eve, he was alone... and I was alone.

I thought of him laying there alone in the dark, left to die. It just didn't set right with me. I wouldn't want to die that way... especially not on Christmas Eve! I held his still warm hand and began talking softly to him, telling him it was okay to let go... head for the light... anything I could think of that would be appropriate to tell a dying man. I dozed off several times, cursing myself for not being more attentive. One time after nodding off I'd have sworn that he squeezed my hand, as if to say "Hey, you were doing fine... don't go to sleep on me now!"

It was around 6:30 AM when a corpsman came in and was startled to find me there with this guy, holding his hand no less. The corpsman had a hint of morbidity in his voice 'till I told him why I'd spent the night, and then he started crying and left us alone.

We packed him off to Da Nang Air Base that day. I have no idea what happened to him after that, but he still had strong vitals when we put him on the medevac bird. I wish I could have remembered his name. I'd promised him I would let his folks know that he wasn't alone on what I believe was his last Christmas Eve, but like so many other promises I'd made to myself or others, once I left 'Nam, all was nearly forgotten. Do we want it this way?

My memories of 'Nam come to me in the form of dreams and nightmares. I find myself trying to escape from a bodybag, or trying to scream out that "I'm not dead!" - but no one ever hears me.

If only I could remember his name....

SEASON'S GREETINGS

Brigadier General Steve Cheney

On Christmas Day 1984 I was a Major with the III Marine Amphibious Force and was assigned as watch officer in the Command Center at Camp Courtney in Okinawa. Sitting there with me were three lance corporals, all of whom were bemoaning the fact that they had drawn Christmas duty. I had volunteered, figuring I'd let some other officer enjoy the day.

The phone suddenly rang and broke the monotony. One of my watchstanders answered, laughed, and then hung up. He turned and said, "Do you believe this? Some nut just called, said he's from the White House, and told me the President wants to talk to us!"

Panic seized me, since I realized that the President *did* sometimes call servicemembers on Christmas… and we may have just hung up on him! I prayed he wouldn't call the Commandant next.

Well the phone rang again, and this time they passed it to me. It *was* the White House Communications Agency, and President Reagan *did* want to talk to the youngest Marine present, so I passed the phone to Lance Corporal Bartlett. The President not only talked to him, but also spoke with his family back in the United States, his Navy brother, and even sent him a tape of the conversation.

It made for a pretty special Christmas present for us all!

O' CHRISTMAS TREE

John E. Lane

In December of 1944 I was serving with the mortar platoon of Golf Company, 2nd Battalion, 25th Marines on the Hawaiian Island of Maui when one of my buddies received a small folding Christmas tree which he set up on an orange crate in the middle of our tent. That was nice, but naturally we couldn't leave it at that.

To make a Christmas scene we cut the camels off packs of Camel cigarettes and hung them on the tree. Then saltine crackers were crushed into a fine powder to make "snow," and a rifle patch was cut into the shape of a star and placed on top.

Another Marine broke out some chocolate covered marshmallows topped with pecans and passed them around as a holiday snack, and it wasn't long before some Christmas carols were being sung. We filled with the spirit of peace on earth, good will toward men.

Unbeknownst to us at that very same moment preparations were going forward for the assault on Iwo Jima, and in a short time we would be boarding ships and sailing for our rendezvous with destiny.

Such is the nature of war.

CHRISTMAS BLUES

Charles A. Robertson

In December of 1966 I was attending disbursing school at Camp Lejeune, North Carolina and feeling "down" - it was my first Christmas away from home, and I was sick with a bad cold and fever.

I began to talk with a classmate named Nick and found out that this was his first time away as well. Misery loves company. Then, out of the blue, he said, "We can't spend Christmas here. Let's go somewhere!"

But where to go? "Washington, D.C.," we thought, since neither of us had ever been there and it was "only" about three hundred miles away. We had just enough money for round-trip tickets and maybe a meal or two, but would have nothing left for a hotel or transportation around Washington. A trip like this made no sense - but the next day was Christmas.

As we left the Jacksonville bus station what had been a rainy day turned to snow, and the farther north we went the heavier it became - and the sicker I got. When we arrived at the Washington terminal it was 2300, dark, and still snowing like crazy. The weather was not a big deal for me since I am from Michigan, but Nicky hailed from South Texas and thought we had mistakenly taken the bus to the Klondike.

We didn't know where we were or where we were going to stay so we just started walking, and right about the time our feet began feeling numb we spotted a Rescue Mission. It cost us two dollars apiece for a room, and I don't know what we would have done if that place hadn't been right where it was or if the price hadn't been so low.

The next morning it was Christmas Day... and all was right with the world.

CHRISTMAS DOWNRANGE

Lex McMahon

Last night my band of brothers assembled for a night of holiday fellowship - Korean barbeque, cigars, and port. As great as the night was there was a void. One of our brothers, Cesar, is currently deployed to Afghanistan for his second tour.

Most of my brothers are part of one gun club or another - Marines, Army, or law enforcement - it does not matter. They've all been in harm's way and fully understand the all too often taken for granted concepts of service and sacrifice.

As we sat around the proverbial campfire

smoking our cigars the war stories invariably came out. We each recounted firefights, ambushes, and assorted near-death experiences such as a RPG attacks while taking a crap.

During the course of our conversation my thoughts kept wandering back to Cesar. What was he doing? Had he been in any significant contact yet? How would he and his Marines celebrate Christmas?

After a great night of testosterone enhancement with my brothers and with Cesar still very much on my mind I headed home to the comfort, love and safety of my family.

Before I went to bed I jumped online to send an email to Cesar to wish him a happy holiday and to let him know that our group of friends had gotten together for dinner and had saluted his sacrifice. Cesar happened to be online, and promptly responded with this message:

"Yesterday was a bad day. Three medvacs, one double amputation... I'm okay, and so are my guys."

As I read Cesar's email I was reminded that the amazing night I had just shared with friends and the holidays I was about to enjoy with family were a gift. A gift paid for by the service and sacrifice of Cesar and his Marines and the other members of our military.

Know this - as you enjoy your holidays with friends and family, you have been afforded this luxury through the service and sacrifice of others. Somewhere in Iraq and Afghanistan a Marine, Soldier, Sailor, or Airman is living in a dirt hole, has not eaten for days, has not bathed in weeks, and is fighting for his life and the life of his brothers-in-arms.

So as you listen to Christmas music, drink eggnog, eat too much food, open presents, and spend time with your loved ones - take a moment to pray for those that have given you the gift of freedom which allows you the opportunity to do all these amazing things. Pray for Cesar.

SEMPER FIDO

Lieutenant Colonel Jay Kopelman
with Melinda Roth

I remember being exhausted. The tiredness weighed more heavily on me than my sixty-pound rucksack. As I walked through the door of our command post in northwest Fallujah after four days of dodging sniper fire and sleeping on the ground, all I could think about was sleep.

That's when I first saw Lava. A sudden flash of something rolled toward me out of nowhere, shooting so much adrenaline into my wiring that I jumped back and slammed into a wall. A ball of fur

Christmas in the Corps

skidded across the floor, halted at my boots, and whirled in circles around me with the torque of a windup toy. Though I could see it was only a puppy, I reached for my rifle and yelled.

It was November 2004. In the days before our march into Fallujah, U.S. warplanes had pounded the Iraqi city with cannon fire, rockets and bombs. The bombardment was so spectacular that I - and the other 10,000 Marines waiting on the outskirts - doubted anyone would live through it. But plenty managed. Now sniper fire came from nowhere, like screams from ghosts.

At the sound of my voice the puppy looked up at me, raised his tail and started growling this baby-dog version of "I am about to kick your butt." Then he let loose with tiny war cries - roo-roo-roo-rooo - as he bounced up and down on stiff legs.

"Hey," I said, bending down. "Hey. Calm down!"

There was fear in his eyes despite the bravado. As I held my hand out toward him, he stopped barking. He sniffed around a little, which surprised me until I noticed how filthy my hands were after almost a week of not washing. He was smelling the dirt and death on my skin.

I leaned forward, but he tore off down the hall.

92

"Hey, come back!"

The puppy looked back at me, ears high, pink tongue hanging out sideways from his mouth. I realized he wanted me to chase him. He was giving me the "I was never afraid of you" routine. So I scooped up the little guy. He squirmed and lapped at my face, which was blackened from explosive residue, soot from bombed-out buildings, and dust from hitting the ground. "Where'd you come from?" I said.

The puppy acted like he had just jumped out from under the Christmas tree, but meanwhile I called my 'cool' to attention. It's not allowed, Kopelman. Marines letting down their guard and getting friendly with the locals - whether pretty girls, little kids, or cute furry mammals - wasn't allowed. But he kept squirming and wiggling, and I liked the way he felt in my hands. I liked not caring about getting home or staying alive, and not feeling warped as a human being because I was fighting in a war.

Born in Pittsburgh and a graduate of the University of Miami, I'd been a Marine since 1992, when I transferred from the Navy. Now in my second deployment to Iraq, I was looking at a starving five-week-old outlaw. Members of the First

Battalion, Third Marines - called the Lava Dogs for the jagged pumice they'd trained on back in Hawaii - said they'd found the pup at the compound when they stormed it about a week ago. He was still with them because they didn't know what else to do with him. Their choices were to put the little guy out on the street, execute him, or ignore him as he slowly died in the corner. The excuses they gave me were as follows:

"Not me, man, no way."

"Not worth the ammo."

"I ain't some kind of sicko, man."

In other words, warriors, *yes*... puppy killers, *no*.

They named him Lava. The newest grunt was treated for fleas with kerosene, dewormed with chewing tobacco, and pumped full of MREs. Officially called Meals Ready to Eat, but unofficially called Meals Rejected by Everyone, MREs were tri-laminate pouches containing exactly 1,200 calories of food. Lava quickly learned how to tear open pouches which were designed to have a shelf life of three years and withstand parachute drops of 1,250 feet or more.

The best part was how these Marines, these elite, well-oiled machines of war who in theory could kill another human being in a hundred unique ways, became mere mortals in the presence of a tiny mammal. I was shocked to hear a weird, misty tone in my fellow Marines' voices, a weird, misty look in their eyes, and weird, misty words that ended with "ee."

"You're a brave little toughee. Are you our brave little toughee? You're a brave little toughee, yesssirree."

The Marines bragged about how he attacked their boots, slept in their helmets and gnawed on all the wires from journalists' satellite phones up on the roof.

"Did anyone feed Lava this morning?" someone yelled out.

"I did!" came back from every single guy in the room.

He was always chasing something, chewing something, spinning head-on into something. He stalked shadows, dust balls and balled-up pieces of paper. He could drag a flak jacket all the way across the floor. But you couldn't yell at him. Even though you were an elite, well-oiled machine of war, you'd be considered a freak if you yelled at a puppy. So he was kept warm and completely pampered.

By the time I came around, he already knew the two most important rules of boot camp: You don't chew on bullets, and you only pee outside. Lava gave the Marines something to be responsible for above and beyond protecting their country, and getting their brains blown out - or worse - in the process. He gave them a routine. And somehow, I became part of it.

Every morning we fed Lava and then piled out of the house to various posts across the city. Some Marines patrolled the streets. Some cleared buildings looking for weapons. Some got killed. Me, I supervised three wide-eyed Iraqi soldiers who, in their new, U.S.-issued chocolate-chip cammies, waved their rifles around as if clearing away spider webs. They were untrained, out of shape and terrified, these members of the Iraqi Armed Forces, coaxed by the United States to help root out insurgents.

At night we all gathered back at the compound where we covered the windows with blankets and sandbags, cleaned our weapons, and made sure Lava had dinner. After that, we would bed down and review the day's events.

"We found a weapons cache..."

"Yeah, well, we got caught in the alley..."

"Yeah, well, we had to transport wounded and then we got hit..."

As we talked, Lava would paw through our blankets. Then he would sit between my crossed legs and stare out at everyone.

As I untied my boots, Lava bit at the laces. As I pulled a boot off, he grabbed hold and tugged. I tugged back. The dog growled. I growled back. "Hey, what's with this puppy anyway?" I asked. "What are you guys planning on doing with him?" No one answered me.

Lava crawled out of my lap and turned a few circles, flopped down and fell asleep with his nose buried in my empty boot.

Like everything else in Fallujah then, nothing but the immediate was really worth thinking about. But when a puppy picked my boots to fall asleep in, I started wondering how he'd die. Especially when I knew I'd be leaving the compound soon and heading for Camp Fallujah about twelve miles away. In February I'd be leaving Iraq for good, and returning home to California.

I just knew the little guy was going to die. This one won't make it because he's too damned cute. As a lieutenant colonel, I also knew military rules as well as anyone, and every time I picked Lava up they darted across my brain like flares: Prohibited activities for service members under General Order 1-A included adopting as pets or mascots, or caring for or feeding, any type of domestic or wild animals. The order was taken pretty seriously. The military didn't want anything like compassion messing things up. Our job was to shoot the enemy, period.

Most nights Lava slept on the roof of the compound with a group of Marines, but once the weather turned colder he came inside. He looked wide-eyed and cute, all paws, snuffles and innocence - but in reality, he wasn't innocent at all. I personally saw the little monster destroy several maps, one cell phone, five pillows and some grunt's only pair of socks.

One morning I woke up and found Lava sitting near my sleeping bag, staring at me, with his left ear flapped forward and the remains of a toothpaste tube stuffed in his mouth. "Morning," I said. He replied with a minty belch.

Another time, I woke up to see his entire front end stuffed into one of my boots, his butt and back legs draped out over the side. He wasn't moving. I thought he was dead - probably from all of those MREs. "Oh, no!" I said, cursing. But when he heard my voice, his tail started wagging like a wind-kissed flag. I decided that from then on he wasn't eating noodles, biscuits or beans in butter sauce. No more toothpaste. Only meat.

And then another morning, I thought someone had short-sheeted my sleeping bag because I couldn't push my feet to the end. It was Lava, who'd somehow managed to crawl in during the night and curl up at the bottom in a ball.

I pulled the dog up under my chin. He snorted and snuffled, and I scratched his ears. "What's going to happen to you once we leave here, little guy?"

The puppy thumped his tail on my chest, and I realized I could no longer sleep at night unless that little fur ball was nestled up against me. Though from day one Lava had been a group project, I was now considering him my own. I made his safety and well-being my mission.

I started calling friends and family, telling them about Lava and asking for help. At first I thought the silences on the other end were the usual international lags on a cell-phone call, but I soon realized my friends back home were trying to put the word "puppy" into the context of war.

When I called one of my best buddies back in San Diego, Eric Luna, and asked him if he knew how to get a dog out of Iraq, I heard nothing for a long time aside from some static. "Hey, Easy E, you still there?" I said.

"Yeah, man, I'm here. What did you just say?"

"Puppy. I have a puppy. Can you help me figure out how to get him out?"

Eric collected his wits. "Sure, man. Yeah, anything you want."

I returned to the main base with Lava on Thanksgiving Day in a Humvee - which, after serial bombardments, firefights and crashes, looked more like a secondhand stock car. Lava loved the loud trip. He perched on my lap and drooled. Once safely at Camp Fallujah, I spoke to the military dog handlers. Their working dogs made up an elite unit that out-specialized any weaponry or high-tech mapping systems the U.S. armed forces possessed.

When I asked if Lava could hide out in one of their kennels, the handlers shook their heads. "Can't help you, sir." They said that the closest military vet who could give Lava vaccinations worked in

Baghdad - some forty treacherous miles away. They doubted he'd be able to help. They wished me luck though, and gave me what I suspected was some very expensive dog food.

When I contacted the military vet in Baghdad he respectfully reiterated General Order 1-A, adding that diseases such as leishmaniasis, hydatid disease and rabies were common among stray dogs in Iraq. "My apparent lack of concern isn't due to not caring," he wrote. "I'm simply following orders."

Well, shoot. But I wasn't about to stop there. I'd already snuck Lava into the officers' building, where he slept with me on a cot. On the computer, I was Googling anything I could think of - including "puppy passport," and "help Marine help puppy." I felt frantic about Lava's fate. Yes, I was a Marine, brave to the point of insanity - but I'd be damned if I was going to let anyone shoot my puppy.

For most of January and February 2005 I worked at the Joint Task Force in Balad, replacing a lieutenant colonel there. I had great accommodations - a trailer with a real bed, a refrigerator, a wall locker. We also had a gym and plasma TVs in our command center. It might have been a great mission, except I worried about Lava. I knew he was safe with the Marines back at Camp Fallujah, but I was trying to save his life.

For a while Corporal Matt Hammond watched him, even building him a little plywood hooch, which the guys filled with toys and blankets and hid in the commanding general's personal security detachment, the last place on earth anyone would think to look. Then we came up with a plan to get Lava to Baghdad, where he would be vaccinated. The guys managed to convoy there, and at a prearranged time and place handed Lava off to journalist Anne Garrels, whom I'd become friendly with and who had promised, by e-mail, to watch him for a few weeks at her National Public Radio (NPR) compound.

The hand-off was a bit of an ordeal, I heard later. Matt struggled to remain emotionless, while Anne grabbed Lava and left. Lava didn't have a collar or a leash, so she had to carry the now-large puppy back to the car. Luckily her Iraqi driver didn't object, as most Iraqis did not like dogs. When I read Anne's e-mail from Baghdad, not even Patton's presence could have kept my tears from flowing. "Just to confirm that Lava is safely with me..."

Was I a gutless wimp? Maybe.

Anne would e-mail me with updates whenever she could.

"Lava is happy."

"He's incredibly affectionate."

"He sits beautifully."

Meanwhile a man she knew in Iraq, someone I'll call "Sam" to protect his identity, managed to locate a vet and get Lava all his shots and proper documentation. Before long Anne had to leave Baghdad, while I was assigned to patrol the Syrian border until leaving for the States. By now I had learned about Ken Licklider, who owned Vohne Liche Kennels in Indiana. He was a former U.S. Air Force police-dog handler who trained dogs for search-and-seizure work, and many of his dogs were used by the military to sniff out bombs in Iraq. There was a chance Lava could fly out with Ken's dogs and handlers to the United States. "It means putting Lava on a transport with them," John Van Zante told me.

John, of the Helen Woodward Animal Center in California, and Kris Parlett, with the Iams dog food company, were my link to Ken. Iams had even offered to pay all the transport costs. Now we just had to sneak Lava out of the Red Zone in Baghdad, where he was hiding with journalists, to the military base in the Green Zone, the walled center of the city. John and Kris would take it from there. Me, by e-mail: "Thanks, John."

John: "We may actually put Lava on a plane. I hope this is it!"

Then, a worry. The kennel's overseas program coordinator: "Can you confirm that Lava has all his health and shot papers in order? Recently we ran into a problem with one of our dogs, and the military vet would not allow the dog to leave the country for an extra thirty days. I don't want that to happen to Lava." Neither did I! On top of that, I was leaving soon.

Sure enough in early March I left Iraq, spent three days in a tent in Kuwait, and then flew to Shannon, Ireland. I was on my way home, but all I could think about as I drank pints with a bunch of other Marines was this - I just didn't see Lava making it to California to be with me. The plan to fly him out seemed too easy. You only get so much luck, my thinking went.

But as the weeks passed, the plan was cemented. In the Green Zone, "David Mack" (not his real name) reviewed Lava's documentation, including an international health certificate for live animals. Security around the Green Zone was cinched tighter than usual after reports of "irregularities" with the Iraqi elections. Demonstrations raged, and mortars were launched.

At the NPR compound in the Red Zone Lava was smuggled into a vehicle with a cameraman, since no animals were allowed to pass through. The vehicle drove to the first checkpoint. Sam waved goodbye. More mortar rounds were launched into the Green Zone. I sat at home in California and waited for an e-mail. And paced. And worried.

The vehicle sped through the dangerous streets, inching toward the checkpoint line. The driver stared forward. The cameraman counted rolls in the coiled barbed wire outside his window.

A bomb dog circled the vehicle as a guard reached through the window to check the cameraman's pass. The pass was good. It was the bomb dog's possible detection of Lava that was so threatening. But he was in search of only one thing, and when he didn't find it he was off to the next vehicle. The guard scanned the pass and waved them into the Green Zone where, at that moment, the Iraqi government extended the country's emergency state by an additional thirty days. All of us waited. I paced some more.

Iraqi police patrolling the parade ground watched a vehicle trailing dust approach a location in the Green Zone and stop. They watched one man get out and shake hands with another, watched the two men exchange papers, watched a dog jump out of the car. They approached the vehicle and asked to see the papers. What was the dog's purpose?

"He's a working bomb dog," one of the men said. "I'm taking him back to my compound." They examined the papers, the dog, the man's face.

A motorcade then sped to Baghdad International Airport. One vehicle contained David, Lava in a crate, other people, and gunmen in bulletproof vests who guarded the doors and windows. The vehicles zoomed along on a highway where twelve people had been killed by bombs in the last month.

Finally my dog arrived at the tarmac near a truck loaded with gear. "This is Lava," David told Brad Ridenour, a dog handler for Vohne Liche Kennels and another vital link in the chain. Soon after, I received a new e-mail.

I stared. I opened it and read. "As of 1600 hours," it said, "Lava is out of the country." For the second time in my adult life, I broke down and cried.

Brad flew with two other dog handlers to Amman, Jordan where they passed through customs. They spent the night in a hotel in Amman, while the dogs were kept in an underground garage. As a result, Brad spent most of the night down there. Lava bounced around and wanted to play.

In the morning the dog handlers were taken to Royal Jordanian, which would fly them to Chicago's O'Hare airport. Ken Licklider, meanwhile, drove to O'Hare where he met up with John, Kris and others. They waited in the baggage area. Finally, Lava's crate came through.

John later explained, "That's when the dam just broke." He told me how he rushed Lava outside and exclaimed, "His first pee on American soil!"

And about Lava's behavior once they got to the hotel room, which John described as "Running around and around the room in circles. Wow."

And then John was finally calling me and saying, "He's here. He's safe. He's an American dog!" John, Kris and Lava flew into San Diego the next day.

Surrounded by the media, I waited at the Helen Woodward Animal Center. Reporters asked me how

I felt, but before I could answer the airport van pulled up. I could see Lava through the window, see how big he'd gotten. I saw the same face, the same goofy look in his eyes, the tongue hanging out.

When Lava hopped down he stopped and stared at all the reporters, then turned toward me, and I looked a little above his head. That way I didn't see the recognition cross his face, didn't see the past and future connect in his eyes. Because if I did I knew I'd lose it then and there, and none of my comrades in the U.S. Marine Corps would ever speak to me again.

I'd wanted him to be alive. I wanted to know he was breathing and leaping after dust balls. If he was alive, then he would make it here to California and run on the beach and chase the mailman instead of strangers with guns. I'd wanted him to be alive almost more than anything I could think of.

Now Lava was headed my way. Fast. As fast as his legs could carry him. As I bent down to deflect the crash, that's when I saw the look in his eyes. It was an older version of the look he gave me when I first spotted him that day in Iraq – "I am going to kick your butt."

Film footage showed a dog barreling toward a

well-composed Marine in uniform who bent down, caught the dog in mid-leap, stood up, and turned circles with his face buried in the dog's fur. Lava was safe. He was home.

CHRISTMAS GIVING

Trina Sonnenberg

About seventeen years ago I was spending Christmas as a newly divorced, single mother. I was twenty-six-years-old, and my son was three. I worked two jobs just to afford to live in a housing project in a very scary neighborhood. I had no Christmas tree, I had no money, I had no gifts for my little boy. Not one! Talk about depressing…

But that changed. Thank God for the U.S. Marine Corps, and a very special friend with a very special family.

That year I stood in line for several hours, waiting

to pick out a gift for my son. I was not in a mall or toy store - I was in a public library, standing in line to receive a gift from Toys for Tots. I was happy to stand there, no matter how long it might take.

I met some terrific people in that line. I heard stories of divorce, of unemployment and of poverty, but everyone I spoke with was happy. No one seemed to be willing to let their financial circumstances ruin the Christmas Season for them. Having walked in their shoes, feeling wretched, I left feeling on top of the world.

When I got home I pulled a letter out of the mailbox, sent from a long time friend and mentor. Inside the envelope was a letter and another, smaller envelope. The smaller envelope contained a check and $200 worth of Toys R Us "Jeffrey" dollars. The letter explained that every year, rather than buying gifts for each other, they pool their money and send it to someone who needs it. She knew I'd gotten divorced, and what I was going through at that point in time, so that year her family chose me.

The whole thing brought me to tears. It also brought me to my knees in thanks. Just as I was drying my eyes, a knock came on the door. The guy I'd been dating was standing in the hallway, with a plastic Christmas tree he'd found discarded.

I then realized that just when I thought I had nothing, I had everything. I had people who cared about me. I had toys for my little boy, and a Christmas tree to put them under. Everything I had, I had because people cared about me.

Christmas is about giving. The people that I met that day gave me a new attitude. They taught me something. They shared their cheer with me, and accepted me even though they didn't know me. The Toys for Tots donors gave me a gift to give my son, and the Marine volunteers gave me their time to see that I got that gift. My friend and her family gave me their love. Instead of giving to each other, they gave to my boy and me.

Christmas isn't about who can rack up the highest credit card bill. It's about making someone else happy - and gifts of yourself mean more and are remembered longer than anything you can get from a store.

A SON'S CHOICE

Roger W. Hoskins

It's a belated Christmas story that started out as a mother's nightmare, as gift-wrapped by her son.

"He said he was quitting Modesto Junior College to join the Marines," Tina Barter said, recalling that day in their Modesto home.

Mom tried to reason with son Jason. "I told him he didn't need to go," she said. "Somebody else's child could serve."

Jason Barter was not swayed by his mother's tears or reasoning. He signed on the dotted line, and soon would be on his way to the Marine Corps

Recruit Depot in San Diego and a war in the desert.

His mother went to bed - and stayed there.

When asked recently what dark thoughts overpowered her then, Tina Barter's face clouded up. She was mute. She refused even to consider articulating feelings. She still was afraid to give any life to a dark fate for her son, who was by then on his second tour of duty.

Tanya Breitenbach only offered that her sister feared the worst, and Tina Barter barely nodded her head in agreement.

But how did she finally get out of bed?

"A good friend came over," recalled Barter. "She got in bed with me and said she was either staying or we were getting up together."

And so she would start to become a full-blooded American military mom.

Today, there is a Marine portrait of Jason Barter in the family's living room. A Marine Corps scrapbook overflows with pictures and mementos. And Barter and her husband Jay were in San Diego when their son graduated from basic training and earned the title U.S. Marine in 2003.

As a reward for finishing boot camp, they took Jason to Hawaii before he was deployed.

During the ongoing war in Iraq, these loving parents waited on pins and needles as their son manned a machine gun on a small armored vehicle.

"This is the damage the vehicle took when it got hit," she said, pointing to a photograph of the caved-in front bumper and hood. "They towed it in and fixed it themselves and went back out."

In another snapshot, the metal hood and fender have been replaced with plywood.

Eventually, Tina Barter gave up her vigil watching CNN and Internet news. She started coming out of her shell. She got a job. Then she started attending meetings of Blue Star Mothers and Families - although reluctantly at first.

"Jennifer (Tyson) invited me, but I couldn't," Barter said. "I wrote Jason about it, and he ordered me to go. He said 'That's the mother of a fallen Marine (Michael D. Anderson Jr.). You *have* to go.'"

She learned to cope. She withstood his first tour of duty in Iraq from March through September of 2004. She even got through a battle in al-Anbar province where twelve Marines died. Her son was there, too.

The next year offered a new challenge for the

Barters. After his first tour, Barter made it home for Christmas. With him and daughter Danielle, the family always had been complete for the holidays.

"Either Jason came home, or we went to him (in San Diego)," explained his mother.

But this Christmas, part of their hearts and home was missing. So they put Christmas in the mail.

Barter was on his second deployment to Iraq. He shared a barracks with six other Marines in Ramadi, where eighty Sunni police recruits had been killed by a bomb a few weeks before. Just days later a mortar shell dropped not far from Barter's office. It's a cause for worry and concern for any parent.

Facing Christmas without her son, Tina Barter decided to do something about it. She enlisted her family, her husband's corporation, a few hotels, and many friends.

Then she mailed the Barter Christmas to her son, piece by piece. Everything was sent - except the family - and she wrote her son not to open any boxes before Christmas.

The one exception was the first box, which contained two Christmas trees, decorations and some inflatable holiday decorations. Barter opened that one to the dismay of some Marine Scrooges.

"What the hell is that, Corporal Barter?" asked a skeptical sergeant. "It's only Thanksgiving!"

That, Corporal Barter replied, is the Barter way.

"That's the way we always do it at home."

When Christmas came Barter opened more than a half-dozen containers, each containing dozens more wrapped presents, and shared his Santa windfall with his Marine comrades.

There were cookies, liqueur-filled chocolates, electronic and board games, balloons, a soccer ball, books and much, much more. There was enough to make many Marines merry, and several cry.

Barter said to his mother, "It's not the best Christmas I've ever had, but considering where I am and who I'm with, it's the best Christmas I could have hoped for." He added to a reporter, "It's been the best part of this deployment."

His mother admitted being a changed woman since that day when her son announced he was joining the Marines. Other people's children matter more than she ever dreamed.

"You cry for everyone," she said. "It's almost as if they are all your own. Whenever I get something for Jason now, I always buy seven."

SANTA'S SLEIGH

This week our daughter Cathy was traveling home from college to be with us for Christmas. When she checked in at the airport in College Station, where she is a student at Texas A&M, she found they had canceled her flight due to mechanical problems with the aircraft. Because of the Christmas holiday, available seats on other flights were almost impossible to find.

After waiting many hours and not finding any seats on other flights, things were not looking good. Alone in the airport (with the exception of workers) she didn't know if she was going to make it home to

118

see us. After sitting on a bench and crying in frustration an older gentleman approached her and asked what was wrong. She explained the situation. They chatted a while, and the gentleman explained he was visiting A&M to see his grandson get commissioned in the Marines.

About that time, they both noticed an old T-28 aircraft taking off from the airport. Cathy explained to the gentleman that her father had flown those as well as helicopters many years ago. The gentleman asked, "In what branch of service did your father serve?"

She proudly said, "The Marines, sir!" Our daughter went on to explain that her mother was also a Marine, and that she even had her very own Gunny who helped raise her and who ultimately retired as a Sergeant Major. She continued on with her story, and sadly told the gentleman that we had to bury "her Gunny" during Thanksgiving.

The older gentleman said, "Young lady, maybe I can help. Where are you headed?" Cathy told him first to Houston with a connection on to Birmingham. The gentleman got up and went over to speak with the gate agent and came back a short time later. He said, "If it is okay with you, you can fly with me to Houston and make your connection."

Not comfortable with "getting a ride from a stranger," Cathy was a bit hesitant. About that time an aircraft taxied up and parked on the tarmac in front of the terminal - and on its side were the words "United States Marine Corps." The gentleman said, "That's my ride, and I can get you to Houston. Would like to come along?"

Cathy quickly responded, "Yes sir!" and off they went. When boarding, Cathy noticed several Marine officers were already in the aircraft, and upon landing in Houston the older gentleman made sure she was escorted by two Marine Lieutenants in dress blues through a private security screening and all the way down to her gate - while catching the eye of everyone in the airport.

She made it home Monday night with a Christmas story she will remember for the rest of her life, and no, the older gentleman *wasn't* St. Nick. He was General Al Gray, former Commandant of the Marine Corps, and still taking care of one of his Marine Corps family.

"THEY CAME IN PEACE"
Christmas in Beirut

Brian Lukas

It's the Christmas season, and every time about this time of year my eyes turn to a specific piece of art positioned just above my desk. There is an inscription I wrote below it which reads:

"Pieces of shredded uniforms littered the branches of the trees; the top of an ammunition can wedged tightly in the bark. A crater in the earth became the resting place for the cinder blocks that once housed the 241 Marines. They died here. 'They Came In Peace.' They were the sons of mothers and fathers, husbands of wives, and dads to their little girls and boys. My wife, my son, and my daughter were home this Christmas day. Jeffrey and Jessica, you were not aware of my assignment, but Mom knew, and she cried. Love, Dad."

Twenty-five years ago Marines were spending Christmas away from their families and loved ones. It was a lonely time - a time that our military personnel are experiencing today, away from family and friends in distant places and distant conflicts.

But this was Beirut in 1983, in another time, and another place of conflict. Civil war had erupted in Lebanon in 1975, the result of clashes between Christians and Muslims and the various political groups of the region. In 1983 the United Nations dispatched a multinational peacekeeping force, including U.S. Marines, to Beirut.

On October 23, 1983 a truck loaded with explosives crashed into the 24th Marine Amphibious Unit Headquarters compound, killing 241 Marines. They seem to have gotten lost in the history books.

In December of 1982 myself, Angela Hill, Garland Robinette and editorialist Phil Johnson had traveled to Beirut to cover Louisiana Marines stationed there at Christmas time.

Christmas was just a few weeks away. It was a time before portable satellite uplinks and the Internet, so we carried videotaped messages from the Marines' families back in the United States. Our ambitious itinerary also included production of a documentary about this war-torn area, but as fighting between the various factions escalated that idea was abandoned. Armed militias set up roadblocks in various sections of Beirut, and the Islamic Jihad decided to add another element to its arsenal of terror and brutality - kidnapping Westerners.

I kept journal entries of the tense times there, excerpted here:

At the same time that I arrived in Beirut, the French Embassy was hit by a car bomb, with twenty people killed. Later that night, a bomb-laden truck blasted a French military base. Ten French soldiers were killed and twenty-three were hurt. The explosion lit up the whole area.

Terror - it is sheer terror. I can see it on the faces of the residents who walk cautiously in the streets.

Here in Beirut teenagers carry assault rifles, mainly M-16s. On the streets women cradle their children tightly in their arms, begging any Westerners for help.

The city smells like death. There is a stench of rotting corpses and smoldering trash strewn about from buildings destroyed by the fighting in the streets. To realize the inhumanity of war, you have to look deep in the faces of the civilian population. Then, if you dare, look deep into their eyes. There you will find the horror of war absorbed deep within the soul.

I look into many eyes here in Beirut. In the eyes of the young Marines, I can see the uneasy and uncomfortable situation they are in. The U.S. Marines' position at the Beirut International Airport keeps them under daily sniper and artillery attack.

I remembered I was in Washington D.C. for a White House press function when many of these same Marines from the 22nd Marine Amphibious Unit invaded Grenada, a tiny island in the Caribbean. Now, I was there in hell with them.

The Marines, politically, were not invaders but so-called "welcome guests," strategically placed in Lebanon on a peacekeeping mission with the French and Italians as part of a multinational force. Our Marine contact, Captain Dennis Brooks - the Marine public-information officer on the base - was always "spring-loaded to say yes." He remarked that the various militias near the Marine positions used their tanks like small arms fire: They quickly maneuvered them into position, released a shell, maneuvered back quickly, and then repeated the operation. Maximum destruction, I thought to myself.

Total destruction was evident when we passed the Sabra and Shatila refugee camps - hundreds, perhaps thousands of Palestinians were killed here. Men, women, and children. Our driver remarked, solemnly, that they were executed.

The refugee camps are now leveled, nothing remains, and where the victims of this civil war sought relief from the terror of war, only the bare reddish-brown earth remains visible from the nearby dusty road. Their graves are not even marked. It is as if they were never born. It was as if Christmas had never come at all.

LIKE AN ANGEL

Elizabeth Westfall Flynn

When I was a kid I worshipped my big brother, Kemper. He was a loyal friend, someone who always faced down a bully, a protector of his three sisters. And he was cool - he did exciting things. When my parents went out of town, he had parties so big it looked like the world had been invited. Everybody loved him - but trouble knew where to find him too.

In 1967 he joined the Marines and fought in Vietnam. By the tender age of twenty, he had witnessed the decimation of his platoon.

When he finally came home, he was different. Quiet, and not interested in the homecoming party my parents wanted to throw for him. Not even excited about the '69 VW Bug they gave him tied up in a bow. He spent the next few years of his life trying to adjust.

But he never did. In 1977 he killed himself, leaving a note asking forgiveness. When his wallet was returned to us, it reeked of exhaust fumes. His death tore my family apart. My parents divorced, and my own heart was broken.

Then on a brisk sunny day a week before Christmas last year, I was out shopping and called home from my car to check on my son. "Mom, some woman phoned and said she was hired by the court to find you. It has to do with your brother."

An old forgotten bank account? I wondered. I called her immediately and was connected with a woman who said she was a confidential intermediary. "I have reason to believe," she said, "you are the biological relative of a female born October 21, 1965 who is seeking medical information. Were you aware your brother fathered a child in 1965? Hello?" I was so shocked I couldn't respond.

My brother's girlfriend had become pregnant when they were in high school, and neither Kemper nor my parents had ever told anyone. Now his daughter was looking for us. I sat in the car with my foot on the brake and just cried.

Bonnie Jean Phoenix had a happy childhood. Her parents were loving and nurturing - exactly the kind of adoptive parents Kemper and his frightened girlfriend would have hoped for. But throughout her life Bonnie had a feeling of disconnectedness, and at age thirty-four she decided it was time to solve the mystery of her origin and began the search. It took her three years.

The day she walked into my mother's house, I was stunned. A perfect stranger who was the image of my brother - his nose, his mouth, his blue-green eyes. She looked like an angel standing in the sunlit hallway. He sent her to us, I thought, to love in his place.

I introduced myself, and before I knew it her arms were around me. She brought a box full of pictures of herself as a child - playing with a pet, swinging in a hammock. A child who always stood up for others - a cheery little girl, her face beaming.

In the days and weeks after meeting Bonnie, I

realized a weight was beginning to lift. It was anger I'd had for years and never wanted to admit. I was angry at my brother for committing suicide. My parents' marriage collapsed, and my sisters and I worried that life's battles might be too much for our own sons. For the next twenty-five years, my brother's death and the manner of it haunted us all.

Then Bonnie found us. She is so much like him. She's reminded me of what a good guy he was. She made me believe in happy endings again. She made me forgive him. It was the greatest Christmas gift of all.

A LITTLE KINDNESS

In the snow-covered desert on the Iraq-Syria border, Marine Major Brian Dennis watched a pack of wild dogs circle the mud fort where he and his unit had stopped to rest and eat. The dogs had come for food scraps, and every time Dennis' Border Transition Team passed through the area to patrol the ruined villages on the border the fifteen mongrels were waiting.

The leader of the pack was a gray-and-white German Shepherd-Border Collie mix with a black snout. The Marines called him "Nubs" because sometime during his short, brutal life someone had clipped his ears. Another casualty of war.

Nubs and his pack had shown up every week for three months. Locals considered the dogs nuisances and thieves but Dennis, a thirty-six-year-old F-18 fighter pilot on his second combat tour, developed a friendship with Nubs. This dog was an oddity. Though he was constantly fighting for dominance among the pack, he was also a clown. Around the Americans he jumped, rolled around, and performed antics for food. He reminded Dennis and his Marine comrades of their pets back home.

That cold day in December 2007, as the dogs circled around, something was wrong with Nubs - he was shaking and barely able to stand. Dennis looked closer and saw that there was a gaping puncture wound in the right side of his chest. The fur around it was matted with dried blood, and the wound looked infected. Villagers told Dennis that an Iraqi soldier had stabbed the dog with a screwdriver.

Dennis couldn't stand to see the dog suffer. He located the team corpsman and together they cleaned the wound, applied an antiseptic cream, and gave Nubs an oral antibiotic. By the time they'd finished, temperatures had dipped below freezing. "I didn't think he would make it through the night," Dennis said.

But in the morning, Nubs was still alive - in pain, and staggering - but alive. The team had to leave, so Dennis knelt down to say goodbye. "Stay strong," he whispered to the dog.

Ten days later Dennis' unit was back - and so was Nubs. He was still weak, but the men fed him and played with him as they always had. When this visit was over and the unit once again pulled out, heading for their combat outpost seventy miles to the south, the Marines watched as Nubs, slow but determined, loped after their departing Humvees. He followed them far into the trackless wasteland until the men lost sight of him.

Two days later Dennis was meeting with Iraqi officers at the outpost when one of his men came running in. "You're not going to believe who's outside," the Marine said. Dennis went out expecting anything - except what he actually saw.

There was Nubs. "That's impossible!" he said. The dog had tracked him across seventy miles of frozen desert, braving wolves and militants, to reconnect with the friend who had saved his life.

It was the Christmas season, but you don't really

have holidays in a war zone. Marines may have something special for dinner, and some pause to pray or reflect. Still, the bombs and snipers don't go away.

"You can't take the day off because it's Christmas," Dennis says. "There's still a mission to do." But this year, in this camp, Nubs arrived as a gift for fifty Marines.

Until the top brass found out, that is. Keeping animals is against regulations, and word came down from above to get rid of the dog by "any means necessary." They gave the men four days.

Dennis had to save this mutt. "We'd slept in the same dirt, run around in the same ruins," he says. "I decided that this dog, who had been through war and abuse, was going to live the good life."

So he looked for a way to ship Nubs to America, searched the Internet, and found a couple in San Diego who were willing to care for the dog until his combat tour was over. Family and friends quickly raised $4,000 for a plane ticket, a travel crate, and vaccinations. Nubs left the desert and flew from Jordan, across Europe, on to Chicago, and then to California. A month later, Dennis arrived at Marine Corps Air Station Miramar, just north of San Diego.

When man and dog were reunited, at first Nubs didn't recognize the guy with the wide smile and shaved head. This was not that dirty Marine in armored battle gear. But within minutes the dog jumped into Dennis' arms, leaping up again and again to lick his friend's face.

"If you do something kind for someone, an animal or a person," Dennis says, "he won't forget you." Nubs didn't forget his friend when he returned to Iraq for another tour the following March, either. The two were reunited again in September, and a children's book about their experience which Dennis coauthored, *Nubs: The True Story of a Mutt, a Marine & a Miracle*, has been published.

A little care and concern in the midst of war will not redeem a violent world. The Christmas promise of peace and goodwill, cynics say, is an illusion. But then there are a million small stories, like the story of a Marine and a dog, to remind us that the impossible is always waiting, and straining to become real.

MY SON, MY HERO

Gail Cameron Wescott

Kendall Phelps will never forget the morning when a picture popped up on the computer in his Silver Lake, Kansas high school classroom. There stood Chris, his oldest son, in desert camouflage in front of a bombed-out building in Baghdad. He was holding a sign written on the back of a discarded MRE carton: "DAD, WISH YOU WERE HERE. SEMPER FI!"

Kendall, a retired Marine, rushed across the street to the elementary school where his wife Sherma teaches fourth grade. She had already received the same photo via e-mail and, in a spasm of giddy relief, was printing out copies to post at the church and all over town. Chris had not been heard from since just after the start of Operation Iraqi Freedom a month earlier. Kendall's elation that his son was alright momentarily erased his disappointment over not being in Baghdad himself.

"Kendall is both a Marine and a dad," explained Sherma. "He could not stand it that he was not over there with Chris, fighting side by side, protecting his son."

That winter, against all odds, Master Gunnery Sergeant Kendall Phelps, fifty-eight, and Major Christopher Phelps, thirty-five, deployed to Iraq together for a seven-month tour in the newly formed 5th Civil Affairs Group based in Fallujah. Ever since the five Sullivan brothers from Waterloo, Iowa had tragically died when their cruiser was sunk in World War II, the U.S. military has been reluctant to deploy immediate family members in the same company - and no one can remember a Marine father and son serving together.

The mission of the 5th Civil Affairs Group was to facilitate the reconstruction of Al Anbar province, an area riddled with suicide bombers and insurgents.

Chris, a team leader, worked with the Iraqi police, firefighters and contractors on rebuilding the infrastructure, and his dad used his thirty years of teaching experience to help establish new schools. They both knew it wouldn't be easy. "It was hot and nerve-racking and people were going to die," said Chris. "My buddies say civil affairs is the most dangerous job in the Marine Corps right now, but I also think it's gratifying to get that grid system up and water pumping." He paused, then added, "And I'd be less than truthful if I didn't say there's some comfort going over with your dad."

In January yellow ribbons sprouted outside Chris and Kendall Phelps' Kansas houses - some seventy miles apart - after the two Phelps men loaded their gear into Chris' Oldsmobile '88 and drove across the country to Camp Lejeune in North Carolina. Kendall, the senior enlisted member of the 200-Marine unit, settled into the routine of predawn wake-ups, long chow lines and field-training in freezing weather as if he'd never left the Corps. Neighbors back in Silver Lake thought he was a little nuts – a father of five (three sons, two daughters) and grandfather of six - voluntarily giving up the comforts of home. "When I asked him why," said Chris, "he said, 'I want to make a difference. I'm a Marine.' For me, there were no more questions. I understood. As a Marine, you feel your time is never over."

Kendall was eighteen when he joined the Corps in 1966. Born and raised in Rock Island, Illinois, he grew up in the poor section of town where no one he knew went to college. The military was his way out of a future in a factory manufacturing tractor parts. Riding around one day with friends, he spotted a large "Join the Marines" poster and decided to talk to a recruiter. Seven months later, Kendall was sent to Vietnam. His parting with his own dad still chokes him up in the retelling. "That was the first time I ever saw my father cry," he said.

Working with KC-130 aircraft in the First Marine Aircraft Wing in Vietnam, Kendall moved ammunition and supplies into and the wounded out of combat zones. "When you sweat, sleep, get scared, get mad and cry together, you find out what you're really made of," he said at a 2003 gathering of Marines. His toughest task was loading the body bags of KIAs, some of whom he knew, onto the planes. To this day he is repelled by the smell of wintergreen, which reminds him of a substance used in disinfectant to plug wounds. "I saw a lot of death," Kendall says. "You don't ever get used to it, but you learn to cope with it."

After returning to Illinois in 1968 Kendall met and married Sherma Meek, who caught his eye in a restaurant when he was out with someone else. By the time Chris was born two years later Kendall, once a talented high school clarinet player, had enrolled in the music education program at Topeka's Washburn University, courtesy of the GI Bill. He soon joined his local Marine Reserve unit, training one weekend a month and two weeks in summer at bases across the country. "I love being around Marines," Kendall says simply. "Always have, always will."

In 1977 Kendall was hired to teach music and help coach at the high school in Silver Lake, a town too tiny for a stoplight, just north of Topeka. "It was like moving to Mayberry R.F.D.," he said. "People leave their doors unlocked. We love it here." Sherma took a job teaching at the elementary level, and the Phelps' settled into a brick and vinyl-sided house with a huge silver maple out front just across the street from both schools. The arrangement allowed the two of them to walk to work and guaranteed the home teemed with friends of their kids. "We even put a phone in the garage so boys could call their parents to pick them up after football practice," said Sherma.

Chris can't remember a time when he did not want to be a Marine. "It began around age two," he said. "I wanted to be like my dad, who was the greatest dad in the world." The two were active partners in Boy Scouts. When Chris rose to Eagle Scout, he asked his father to give the speech at his Court of Honor. Kendall also guided his son in track to a high jump record that stood for more than a decade until Chris' younger brother Josh broke it.

When he was seventeen Chris tried to join the Marines' delayed-entry program (a commitment to boot camp within a year), but his dad flatly refused to sign the initial parental form. "You are saying you're willing to put your life in jeopardy," Kendall explained. "You've got to do that *yourself*." The following April, on his eighteenth birthday, Chris signed the documents and left for San Diego and boot camp right after his high school graduation.

That fall, while simultaneously serving as a private in his father's Marine Reserve unit in Topeka, Chris entered the University of Kansas. He earned a bachelor's degree in literature and a master's in administration, but yearned to become an officer. "The esprit de corps just gets in your blood," he said. "Band of brothers is such a cliché - but that's what the Marines are." Just six weeks after his June 1994 wedding to Lisa - a blue-eyed brunette he vowed to marry the first night they met - Chris entered Marine Officer Candidates School.

On the cold December day when Chris was commissioned a second lieutenant, after the parades and ceremonies, his dad waited for him on a street in Quantico. Standing at attention, Master Gunnery Sergeant Kendall Phelps proudly gave his son his first salute. Then, as Chris' wife, mom and younger siblings noisily crowded around the new officer, Kendall slipped behind a building to wipe away his tears. "Marines," he explained, "are not supposed to cry."

It was even tougher to hold back on the day Chris left for his first tour in Iraq in the winter of 2003. By then, Marine Corps regulations had compelled Kendall to retire from the Corps after thirty years of service. Chris had become the commanding officer of his father's former unit, which had just been mobilized as an ammo-resupply platoon. "In combat, that can be a very dangerous place to be," said Kendall.

That morning at his home near Kansas City, Chris said goodbye to his three little boys and his wife, who was pregnant with a fourth. His mother and father joined him for his send-off from Reserve headquarters in Topeka and Kendall, more dad than Marine this time, made no effort to conceal his emotion. The next day, he stepped up his efforts to return to active duty.

Operation Iraqi Freedom unfolded in agonizingly slow motion in Silver Lake. By day, Kendall taught kids to play instruments and coached high school track. In the evenings, he retreated to his house to monitor the coalition forces charging across the Iraqi desert on CNN. Sherma, who couldn't take her eyes off the ticker listing casualties, got accustomed to finding her husband on the sofa, still staring at the television screen at three o'clock in the morning.

Even after Chris returned home his dad, suspecting his son would be sent for another tour, persisted in his efforts to be reactivated. "I know a lot of Marines," he said, "and I called all of them. And I *kept* calling."

His determination was finally rewarded just

before Christmas when he received an e-mail from the commanding officer of the 5th Civil Affairs Group. It read: "Prepare for a long stay in the desert." Chris was recruited for the same mission. Sherma could not believe it. "I just had a strong feeling it was not going to happen," she said quietly. "Then it happened. I knew I'd be a basket case until they returned."

Christmas was as festive as ever. "We didn't talk about what was about to happen," recalls Sherma, "because with six grandkids running around, it's hard to talk about anything." But after the New Year celebrations, the stress escalated.

On the January night before Kendall and Chris left home, Chris said goodbye to his family. His four sons - Tristen, Dalton, Preston and Taigen, gave him a drawing of a blue angel with a halo, saying it would protect him in Iraq. Chris folded the drawing and put it next to a Marine Corps prayer he carries in his pocket. When Tristen confessed he was scared, Chris hugged him and said, "Me too, buddy, me too."

Meanwhile, in Silver Lake, the Phelps' daughters and two younger sons gathered for pizza at the table where they'd shared so many meals. At 5 AM

Sherma, Kendall and son Josh drove down the empty expressway to Chris' house, where the car was already loaded for the trip to Camp Lejeune.

Kendall hugged his wife, and both cried in the predawn darkness. Lisa and Chris kissed and then, pretending he'd forgotten something in the house, Chris raced back inside and left a letter he'd written on the kitchen counter. In part, it read, "Boys, I want you to know that I am a Marine, and I do what I do for you. We live in the greatest country in the world, and because of that your options throughout your life will be limitless. You have those options because there are hundreds of thousands of servicemen and servicewomen willing to protect America and our way of life. I am no different than any of them. I am so very proud of each and every one of you. You all have special God-given talents and you are destined for great things in the future. Remember to always be honest, keep your integrity, speak your mind, and fight for what you believe. I love you all very much! Semper Fidelis, Dad."

Thereafter Lisa and the boys read the letter aloud often, and every night each slept in one of their father's Marine or KU Jayhawk T-shirts. "It made us feel a little better," said Lisa.

While profoundly proud of their men, neither Lisa nor Sherma pretended that it was easy being separated. Said Sherma, "Every time Kendall went to summer training over the years, something would happen at home - a flat tire, a tornado, even a snake in the house. This time, sure enough, a strong wind blew the water heater pilot light out - something that had never, ever happened before." And at Lisa's, the dishwasher broke down, Tristen had to go to the ER with a 103-degree fever, and the vacuum blew up - all within the first three days of Chris' departure.

Lisa did her best to stay strong for her little boys, and with her husband and son both in Fallujah, Sherma sometimes spent sleepless nights trying to stave off worry. "I just prayed that they got back here together safely," she says. "Please. That's all I wanted for Christmas."

CHRISTMAS IN THE CORPS

T'was the night before Christmas,
And all through the Corps.
Not a soul had liberty,
And the troops were all sore.

Yes, every Marine,
Every one in the lot.
Was lying on a rack of nails,
Called a Marine Corps cot.

Christmas in the Corps

When out on the Parade Deck,
There arose such a clatter.
I sprang from my cot,
To see what the hell was the matter.

With bayonet in hand,
I moved stealthily to the door.
And waited to see,
If there was something more.

T'was the Commandant of Marines,
Of this there was no doubt.
Because he wore a poncho,
With the green side out.

He quietly moved from rack to rack,
And carefully inspected each rifle and pack.
To a chosen few he left a 96 chit,
But to the majority he gave a ration of shit.

And as he pulled away in his gold plated tank,
Pulled by ten captains all bucking for rank.
I heard him say, and he said with a shout,
Merry Christmas you bastards, you'll *never* get out!

Christmas in the Corps

www.ingramcontent.com/pod-product-compliance
Lightning Source LLC
Chambersburg PA
CBHW071002040426
42443CB00007B/618